Building Dilapidation and Rejuvenation in Hong Kong

Editor: Andrew Y. T. Leung

Associate Editor: C. Y. Yiu

Building Surveying Division
The Hong Kong Institute of Surveyors

City University of Hong Kong Press

First published 2004
Second printing 2005
Printed in Hong Kong

ISBN 962-937-090-5

Published by
Building Surveying Division
Hong Kong Institute of Surveyors
and
City University of Hong Kong Press

Website: www.cityu.edu.hk/upress
E-mail: upress@cityu.edu.hk

Contents

Foreword

In recent years, the general public of Hong Kong have started to realize the importance of building maintenance. Most people now agree that proper maintenance of buildings is important for maintaining safety, good health and good value for property owners, residents, and investors. People also understand that deficiency in building maintenance may cause major problems to the community as a whole.

However, books covering this important topic in urban planning and development with specific reference to the Hong Kong situation are rare. We are therefore delighted to offer this book as a pioneer with a view to stimulating further studies and discussion on this topic. We hope that the research findings and critical insights presented in this volume will contribute to the improvement of the living environment in Hong Kong, and hence the sustainable development of the city's urban landscape in the long run.

The Building Surveying Division of the Hong Kong Institute of Surveyors is very glad to have the chance to participate in the publication of this book. We sincerely hope that this book will receive attention by building professionals, policy makers, as well as the general public.

Raymond Chan
Chairman
Building Surveying Division
The Hong Kong Institute of Surveyors
February 2004

Preface

Buildings create an environment where people work and live and therefore affect their quality of life. In Hong Kong, as in other major metropolitan cities, buildings define the image of the city and constitute considerable amounts of financial investments by individuals, families and businesses. For all these reasons, it is important to understand the life-cycle of buildings, how they age, and what can be done when they become old. Ageing buildings and the problems and opportunities they present have significant policy ramifications and implications for both the building and social sciences.

As the economy of Hong Kong grows and changes, the number of new buildings also steadily increases. At the same time, more buildings are falling into a dilapidated condition, which further inflates the already plentiful ageing building stock. The demand for maintenance and refurbishment of existing buildings is therefore significant. Most buildings, especially those in the older built-up areas, lack proper repair, and in time they create an unsightly urban environment where the majority of people work and live.

This book provides an understanding of the life-cycle of buildings, how they age, and what can be done when they become old. The contributors are specialists in their respective fields — architecture, building, fire and structural engineering, surveying, economics, psychology, and social work — and their critical insights into the problems and challenges of urban renewal in Hong Kong reflect the scope and scale of the long-term impact of government policy on the living conditions of the territory as a whole. The papers propose new dimensions in evaluating urban renewal and addressing how, through rejuvenation, the value of ageing buildings may be increased. It is useful not only for policy makers, social workers and building professionals, but also for students and researchers in this field.

The introduction gives us an overview on the history and the latest development of building control in Hong Kong. Chapters 1 and 2 identify the existing conditions of buildings and the government's role in building mainten- ance. Various condition surveys of buildings are reported. Chapter 3 reviews the historical development and the current issues of urban renewal and the 'people-centred' mission adopted the Urban Renewal Authority is recommended. Chapter 4 discusses the effects of zoning on housing prices. A development in a Comprehensive Development Area is studied. The authors suggest that urban renewal should involve broad community support and involvement. Chapter 5 estimates the value enhancement by carrying out refurbishment. These 3 chapters provide three different dimensions in discussing redevelopment and refurbish- ment in urban renewal strategy. To understand the human meanings of living in aged buildings and its implications of a review of the existing urban renewal policy, Chapter 6 explores the quality of life of residents in five urban renewal districts. An index is developed to assess the quality of life in the districts. Chapter 7 develops a tool for assessing building performance: the value age index to facilitate the decision between redevelopment and refurbishment. The findings show that the structural and system defects of a building are the most important criteria, and building age is much less important than expected. Chapter 8 provides a systematic framework for the ranking fire risk of existing buildings, while Chapter 9 compares two evaluation methods of fire risk of existing buildings.

Andrew Y. T. Leung
Professor (Chair)
Department of Building and Construction
City University of Hong Kong

Introduction

It is obvious that buildings are the principal objects that building surveyors care for. Buildings are the "patients" of building surveyors, the so-called "building doctors". Their state of health and their prolonged economic lives are a paramount to the sustainability of the built environment. Buildings not only demand the services of professional building surveyors, but also call for the attention of other parties. These include the architects and engineers who design the buildings, the builders and contractors who construct them and the landlords and tenants who own and derive benefit from them. They also include the building managers and maintenance personnel who manage and maintain the buildings, the occupiers and visitors who use and enjoy them, and the passersby and other members of the public who simply admire them and feel proud of them. Buildings symbolize the achievements and accomplishments of the city and deserve the utmost respect from the society at large.

Due to the lack of a sense of belonging among some Hong Kong citizens, it is inevitable that a "couldn't care less" attitude has begun to dominate people's behaviour in this "borrowed place", i.e. Hong Kong. This attitude has grossly undermined the respect due to our buildings. This lack of respect induces a lack of attention and a lack of care for buildings, and has led to their rapid dilapidation and deterioration. When the government makes more and more social benefits available to Hong Kong citizens, people become more dependent on the government and expect, and even demand, that the government should do for them what they should be doing for themselves, particularly in respect of the care and maintenance of buildings. Similarly, when society becomes more affluent, people tend to extend and maximize their privileges in all aspects of social life, even to the extent of abusing the law. This has indeed been the case of the building law in Hong Kong. Extending their accommodation space with unauthorized building structures and unscrupulously altering the buildings to suit their unique life styles are just two of many examples of a general public disregard for the law. On the other hand, the ineffectiveness of the Building Authority in rooting out such contravention is attributable to the rigidity of the building law. The prevailing Buildings Ordinance has been given only piecemeal touch-ups in the past 50 years or so since its enactment in 1955, and the building control regime has remained out of touch with the fast-changing social realities. This fragmentation has been exacerbated since the decentralization in 1982 of the then Public Works Department into a number of autonomous departments. Before this time, the Director of Public Works was not only the Building Authority, but also had authority over land and town planning, highways and civil engineering, as well as water supply and sewage disposal. But since 1982, his comprehensive authority and influence over building development was thence effectively broken up.

In recent years, statutory building control has been undergoing a drastic "cultural change" and the Buildings Department is prepared to "go the extra mile", which is the latest motto of the Department. The Building Authority now accepts and encourages "green" and innovative designs and construction technology. It has re-engineered the approval process for building plans, it is transforming prescriptive design standards to building performance standards, and recently it has introduced a self-certification regime in respect of minor works.

According to proposed the Buildings (Amendment) Bill 2003, what was previously often claimed to be "exempted works" by virtue of section 41(3) of the Buildings Ordinance will be curtailed. Only those cosmetic and insignificant works carried out inside an existing building that do not alter the structural elements of the building and do not carry any load other than their own weight will be exempted from the requirement of prior approval and consent. Any work that does not involve the structure of the building but is not confined to the interior of the building will not be exempted. Nonetheless, the Bill introduces a new category of minor works to cater for minor structural works and non-structural work. The Bill proposes that such works will be exempted from the requirement of prior approval and consent but makes them subject to self-certification.

Such minor works will include alteration works that may or may not involve the structure of the building, or additions of amenities. They may be carried out by minor works contractors, who form a new category of contractors proposed in the Bill. The new legislation will enable the owners or occupiers to improve the quality, to enhance their usefulness, and to extend the economic lives of existing buildings. This is certainly a major change in the building control system that will greatly facilitate the rejuvenation of our dilapidating housing stock.

While the good intention of the minor works control regime is fully appreciated, it is more realistic for the public to expect a system that is simple and effective. Owners and occupiers of buildings must be able to understand their positions unequivocally and must be able to identify and select appropriate contractors and professionals to effect any renovation or rejuvenation of their property. Any ambiguity in understanding or in the interpretation of the legal requirements should be removed and any cumbersome formalities should be obviated.

To ensure the effectiveness of the new legislation, building owners must be fully aware of their liabilities. They should undertake to comply with the Building Regulations and to abide by any requirement or condition the Building Authority may impose. They should appoint a competent registered minor works contractor to carry out the works and to report and certify completion to the Building Authority. In this regard, it is considered more appropriate and helpful for minor works contractors to be registered according to their respective designation and specialization, for example, general building works, drainage works, demolition works, and so on, so that the relevant contractor may be appointed for the specific task in hand.

Before any works are carried out, the owner must notify the Building Authority of his intention to carry out minor works. This will allow the Building Authority to activate his monitoring mechanism and carry out auditing inspections of the minor works as and when necessary. Where the proposed minor works will affect the original design parameter of the building, or will alter the original provisions for fire-fighting, fire-resisting or means of escape, the works should be designed and supervised by an authorized person. To accompany the notice of the commencement of works, plans prepared by an authorized person should be submitted to demonstrate the measures proposed to maintain or reinstate the safety of the building. Similarly, where structural works are involved, plans prepared by a registered structural engineer should be submitted to demonstrate how the structural integrity of the building is to be maintained. Where any change in use is intended, the notification requirements under section 25 of the Buildings Ordinance can be integrated with the notification of the proposed minor works. This will provide an opportunity for the Building Authority to prohibit any unsuitable change in use or any unacceptable minor works, where the circumstances so warrant.

When considering the rejuvenation or revitalization of our existing building stock, one of the many objectives must be towards sustainability. The overall improvement of existing buildings may be handicapped by their inherent built forms and site constraints. However, it is still not impossible to introduce incremental improvements to energy efficiency and operational efficacy, so that both the internal and external environmental quality may be maximized and adverse impacts on the buildings themselves and their neighbourhood may be reduced.

The Building Authority has lately been placing more emphasis on environmental issues and is providing incentives to encourage more healthy and sustainable buildings. This is yet another paradigm shift in the traditional scope of building control. However, whether any improvement can be implemented depends heavily on the voluntary initiatives of developers and building owners and their building management. There is no threshold and no comprehensive standards for such goals and these improvements are inevitably fragmented and may sometimes be introduced at the expense of other values. In order to help raise the general awareness to a common level of expectation, the Buildings Department has commissioned a research project with a view to devising a *Comprehensive Environmental Performance Assessment Scheme (CEPAS)*. The scheme is expected to provide an objective evaluation of all environmental issues relating to both new building development and existing buildings.

This research is currently in the development stage and there is still no conclusion yet as to the system of rating and the strategy of implementation that will be employed. Notwithstanding this, it is understood that CEPAS will be tailored for different stages of the life cycle of a building encompassing the pre-design, design, construction and operation stages. The first three stages will be applicable to new building developments as well as to major renovation projects. The last stage will be a recurrent exercise that can be executed many times throughout the life span of the building. In addition to establishing measurable criteria for environmental performance, CEPAS is also expected to create a labeling scheme to denote the different performance status of the buildings. On the one hand, this will set objectives for environmental quality achievement or enhancement. On the other hand, the quality performance labels will increase the marketability of those buildings. To the consumers and to the general public, these will constitute an assurance of healthiness and sustainability.

While we are looking forward to the eventual enactment of the new minor works regime and the introduction of CEPAS that would facilitate the actual refurbishment and renovation of our housing stock, it is also appropriate to examine the issue from various other perspectives. The following chapters will analyse the overall societal setting circumscribing housing dilapidation and rejuvenation in Hong Kong.

Barnabas H. K. Chung
Chairman
Board of Professional Development
Hong Kong Institute of Surveyors

Acknowledgements

We would like to thank the Building Surveying Division, Hong Kong Institute of Surveyors for their publication grant for this project. We would especially like to thank the authors for their contributions to this publication. We owe a debt of gratitude to Carole Pearce for his careful reading of the entire manuscript. We would like to express our sincere appreciation to Mr. Patrick Kwong, Director of the City University of Hong Kong Press and Dr. Vivian Lee, Editorial Manager, for their valuable ideas and editorial advice. Our thanks also go to the Department of Building and Construction and the Department of Applied Social Studies, City University of Hong Kong for the joint research grant: Ageing Buildings and People (SRG #9010006), which supports Chapters 2, 3, 4, 5, 6 and 7. We also thank the Research Enhancement grant 2003 for inviting Raymond Pong as Visiting Professor at the Department of Applied Social Studies, City University of Hong Kong. Chapter 5 was also supported by the research grant of the Research Grant Council of Hong Kong (RGC Reference Number: HKU702/98E) and the University of Hong Kong Outstanding Young Researcher Award. Chapters 8 and 9 were fully supported by a grant from the Research Grant Council of the Hong Kong (RGC Reference Number: CityU1112/99E).

Andrew Y. T. Leung
Professor (Chair) and Head
Department of Building and Construction
City University of Hong Kong
February 2004

List of Illustrations

Figures

Tables

1

Maintenance of Old Buildings

Kenneth J. K. Chan

Old buildings are part of our heritage, but they can pose serious problems to our society if they are not regularly and properly maintained. This article reviews existing problems associated with the lack of maintenance of aged buildings. It shows that the government has taken a passive role in preventing the deterioration of private buildings and the inadequacy of the law does not help in promoting the maintenance of existing buildings. Other problems that hamper the proper maintenance and repair of buildings are then examined. It is recommended that a new culture in the maintenance and management of our building stock should be established and the government should relinquish its former crisis management role and take a bold and responsive step to legislate for the proper maintenance of private buildings in Hong Kong.

Chapter

Maintenance of Old Buildings

Kenneth J. K. Chan

1 Introduction

Old buildings do not necessarily pose a problem unless they have deteriorated due to lack of care and maintenance. Chan (2000) referred these problematic buildings as "aged buildings". There is no survey on the number of aged buildings in Hong Kong, but the Housing, Planning and Lands Bureau (2003) estimated that there are about 9,300 private buildings in the Metro Area (i.e. Hong Kong Island, Kowloon, Tsuen Wan and Kwai Tsing) which are 30 years' old and above. In ten years' time, the number of buildings over 30 years' old will increase by 50%.

Multi-ownership in high-rise aged buildings is commonly found in Hong Kong. Many of them are not properly managed. Even if a management agent is employed, very often only caretakers are provided. Repairs, rather than maintenance, are carried out on a reactive basis. There is no sinking fund for improvement works. Usually repairs are carried out only when they are ordered by the Building Authority. Owners are willing to spend on maintaining their own units internally and neglect external common areas.

The general awareness of owners of their needs and obligations to maintain their properties is very low. They invariably adopt a "wait-and-see" attitude until problems arise. Such problems of building dilapidation cannot simply be solved by education. Legislation must be enacted to make it a statutory obligation to owners for manage and maintain their buildings and to make them aware of this obligation.

2 Inadequacy of the Laws and the Government's Passive Role

The government has done very little to prevent the deterioration of private buildings in the territory. The bulldozer of economic success has helped solve the problem in the past decade. The plummet of the economy recently, however, has slowed down the urban renewal process. The maintenance of aged buildings becomes much more critical. There is no mandatory building inspection or maintenance scheme so far, and landlords are only implicitly required to maintain their properties. This has never been a line of enforcement by the government although it is the ultimate freehold owner.

Section 26 of the Buildings Ordinance[1] provides that

> "Where in the opinion of the Building Authority any building has been rendered dangerous or liable to become dangerous by … dilapidation, … the Building Authority may by order in writing served on the owner declare such building to be dangerous or liable to become dangerous "

Until more recently, in 1992, Section 26A of the Buildings Ordinance was enacted to cover "dilapidated or defective buildings". The new provision is:

> "Where, on inspection, the Building Authority finds any dilapidation or defect in a building he may by order in writing served on the owner of such building require such works as may be specified in the order to be carried out."

This provision is an addition to the old provision where the Building Authority could act only on buildings that are dangerous or liable to become dangerous. However, the enactment does not help in promoting the maintenance of existing buildings, as it still relies on the inspection of the Building Authority. Furthermore, the Building Management Ordinance[2] was enacted to set a framework for the management of existing buildings. Unfortunately, it cannot promote the sense of responsibility in building maintenance.

The seriousness of building dilapidation can be shown in the following table and figures. Firstly, Table 1.1 shows the increasing trends in the reports received of various building dilapidation signals. Figure 1.1 depicts the high level of the issuance of orders and the receipt of reports on dangerous buildings. However, Figure 1.2 reveals that the rate of compliance to these orders is decreasing. Figures 1.3 and 1.4 report the exponentially increasing rate of order issuance on dangerous advertising signs and unauthorized building works in the past five years. Unfortunately, Figures 1.3 and 1.5 show that the compliance rates to the issued orders on dangerous advertising signs, unauthorized building works and defective drains are extremely low,

Table 1.1   Reports Received about Dangerous Buildings, Hillsides, Advertising Signs, Unauthorized Building Work and Defective Drains

| Year | Dangerous Buildings | Dangerous Hillsides | Dangerous Advertising Signs | Unauthorized Building Work | Defective Drains | | Total Nos. of Reports |
					Industrial Source	Non-industrial Source	
1998	3,851	53	250	12,577	0	296	17,027
1999	4,730	130	614	16,999	0	365	22,811
2000	4,280	71	260	13,911	7	327	18,856
2001	6,671	41	178	12,764	25	527	20,206
2002	5,956	52	135	21,844	20	554	28,561

Source: *Buildings Department* (1998–2002).

Figure 1.1   The Situation of Dangerous Buildings in Hong Kong

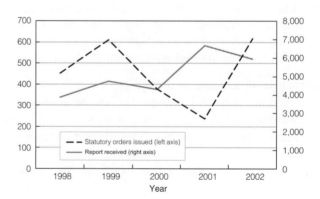

Figure 1.2   Compliance with Statutory Orders on Dangerous Buildings in Hong Kong

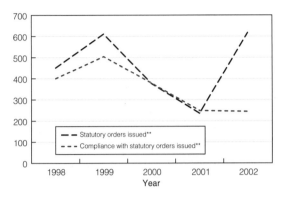

Figure 1.3   Compliance with Statutory Orders on Dangerous Advertising Signs in Hong Kong

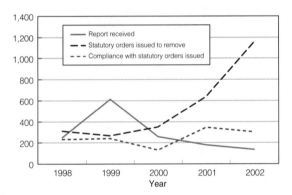

Figure 1.4   Compliance with Statutory Orders on Unauthorized Building Work in Hong Kong

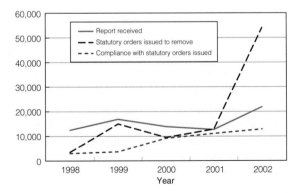

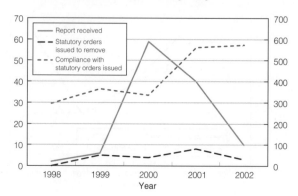

Figure 1.5   Compliance with Statutory Orders on
Defective Drains in Hong Kong

Since the government has taken a *laissez-faire* attitude towards promoting building management and the sense of maintenance responsibility among building owners. This passive attitude has incubated the attitude among owners that there is no problem unless there is specific condemnation from the government.

3 Associated Problems

Other problems that hamper the proper maintenance and repair of buildings are the proliferation of unauthorized building works, the uncontrolled installation of advertising signs, the division of responsibilities and unfair terms in the Deed of Mutual Covenant[3].

In various seminars and workshops on building maintenance in different districts, owners invariably raise the issue of unauthorized building works. Companies and individuals complain of the attitude of some owners who put up unauthorized structures. Many of these unauthorized structures are erected on the exteriors and flat roofs of buildings in ways that often hinder the carrying out of repairs and maintenance to the exterior and drainage system of buildings. Many components of the drainage system are made inaccessible by these unauthorized structures. They have also given rise to disputes among the co-owners. Owners are often unable to deal with such problems without resorting to tedious and expensive legal proceedings. In some instances, the buildings have been altered to such an extent that common areas, such as fire escape routes, are affected.

Similarly, abandoned and dilapidated signs do not only pose a safety hazard to the public, but they also make it difficult for owners to carry out repairs and maintenance to their buildings. They affect the structural safety of the buildings where they are attached to. Many owners are unwilling to pay for the cost of removing these signs. There have also been incidents over unfair terms in the Deed of Mutual Covenant when the first owners or their assignees have retained control of the external wall and have thus put advertising signs on it, contrary to the wishes of other owners.

The problem of dilapidation is further aggravated owing to the division of responsibilities between adjoining owners. In many buildings in Hong Kong the service pipes of a particular unit are buried in the structural elements of another unit that the first owners may not have access to. Problems arising from these pipes affect owners who do not have control over the area where the works must be carried out.

4   Beyond Old Buildings

Sooner the new buildings will catch up and become old in their turn. These new buildings will run into the same problems if nothing is done to guarantee their maintenance and care now. A new culture in the maintenance and management of our building stock must be introduced before it is too late and too expensive to do so. Owners must be put under statutory obligations to manage and maintain their properties properly. Owners of newly occupied buildings must also be asked to contribute towards the future maintenance of the property.

Similarly, we need new legislation to provide for safety and hygiene inspection and planned maintenance. If buildings are not now required to be inspected at prescribed intervals to allow inspection and maintenance/repairs to be carried out, the cost of deferred inspection and repairs when the building is old will be much higher. The risk of breakdown or the failure of individual building elements will then be greatly increased. Society will also have to pay a high cost because of the danger posed by these deteriorating buildings and their unsightly appearance. The proposed legislation should require the government, as a reminder, to place a notice in the Government Gazette at the beginning of each year, publishing a list of buildings requiring general appraisal and requiring a report back to the Building Authority. Owners will have one year to comply with the requirements of the new legislation.

I also appeal to the government to include professional surveyors (building surveying), who are competent to carry out a general appraisal of building conditions, in the Registry of professionals to carry out the inspection task. Registered Professional Surveyors in the Building Surveying Division have been uniquely educated and they are trained and experienced in the care and maintenance of buildings. Their skills consist of the diagnosis of building defects, understanding the performance of buildings in use, maintenance management, repairs and remedial measures and so on, that are indispensable in carrying out the necessary maintenance and repair. Moreover, the staffs of the Buildings Department, who carry out such inspections, may not be Authorized Persons or Registered Structural Engineers as defined in Section 3 of the Buildings Ordinance. I believe the success of the scheme depends on the support of the building owners and the availability of suitably qualified professional persons to participate in its implementation.

One of the problems encountered by building owners in carrying out repairs and maintenance to their properties is their lack of power to ensure compliance by all owners. In most of these cases, the majority of owners have no alternative but to abandon their efforts and leave it to the Buildings Department to act in default. In connection with this new legislation, the government must put in place the infrastructure required to facilitate smooth implementation. The government should revamp the Building Management Ordinance by introducing incentives for the formation of incorporated owners and the introduction of building professionals in the management and maintenance of private properties. Each development will then have a properly organized management corporation consisting of company owners and management/maintenance professionals or registered property management companies. Furthermore, developers should be asked to contribute 0.75% of the proceeds from property sales for setting up a maintenance fund for their properties. This will ensure that the management will start off with a solid footing. It should, however, be borne in mind that some of the properly managed residential estates would opt not to form Incorporated Owners as the current management structures are better fit for their purposes. Certain exemptions should therefore be introduced.

The government should also extend the Improvement Loan Scheme to cover grants to the less wealthy owners to undertake the repairs and maintenance needed by their properties.

The Buildings Department has already banned water pipes from being buried in structural elements. In the planning and design of new buildings, developers must ensure that the property of individual owners can be serviced. Common service lines must be easily accessible. Those belonging to one owner must not pass through any area that belongs to another owner.

Internal service/pipe ducts were in the past too small for proper maintenance to be carried out. When it comes to repairs, new pipes were installed on the exterior and the internal pipes were simply abandoned. The granting by the authority of gross floor area concessions for serviceable pipe ducts was the right direction to take. Access for servicing should be carefully considered and provided for at the design stage. These measures facilitate ongoing maintenance without having to resort to the construction of expensive temporary work access or platforms when works are carried out.

5   Conclusion

As a concerned professional, I call upon the government to relinquish its former crisis management role and take a bold, determined and responsive step to legislate for the proper maintenance of private buildings in Hong Kong and to invest as much on maintenance as has been put into public housing and government buildings. The above suggestions and current related provisions under the Buildings Ordinance should be brought together in a new enactment to be known as the Building (Management Inspection and Maintenance) Regulations to be administered by the Commissioner of Buildings. This will give weight to the issue and give the general public a clear indication of the government's determination to ask building owners to conduct proper maintenance of their buildings.

The Commissioner of Buildings should further embrace all the building activities currently administered by different agencies or departments. Such activities include but are not limited to alterations and additions, the licensing of premises for various trades and uses, the rules concerning defective drainage, control and enforcement of unauthorized building works, as well as fire and building safety. These activities will be dealt with more appropriately and effectively by one authority alone. Only if we act now we will be able to give our next generations a heritage that they can be proud of.

Notes

1. Cap. 123, Laws of Hong Kong

2. Cap. 344, Laws of Hong Kong

3. It is the agreement between the co-owners to regulate their co-ownership of the building and provide for the building's effective management. (Nield, 1997, 399)

References

1. Buildings Department. 1998–2002. *Statistics on Existing Buildings*, Hong Kong: Hong Kong Special Administrative Region Government.

2. Chan, J. K. 2000. Maintenance and Repairs of Old Buildings, *Hong Kong Institute of Surveyors*, June, 4–7.

3. Housing, Planning and Lands Bureau. 2003. Urban Renewal and Buildings. *Homepage of the Bureau, http://www.hplb.gov.hk/eng/policy/urs.htm*, 13 November 2003, Hong Kong Special Administrative Region Government.

4. Nield, S. 1997. *Hong Kong Land Law*, 2nd Edition, Hong Kong: Longman.

2

A Review of Building Conditions in Hong Kong

Andrew Y. T. Leung and C. Y. Yiu

The problem of building dilapidation is exacerbating in Hong Kong due to the building boom in the 1960s. The construction quality of buildings of this vintage has a far-reaching effect on the overall rate of deterioration of the whole city. This chapter first reviews previous technical studies of building qualities of different vintages. It shows that buildings dating from the 1960s are of exceptionally poor quality. Various condition surveys conducted by government departments are then examined. The ageing trend of housing and the status of unauthorized building works are also studied.

Chapter

2

A Review of Building Conditions in Hong Kong

Andrew Y. T. Leung and C. Y. Yiu

1 Introduction

This chapter presents an overview of the current status of private housing stock in Hong Kong. We start by presenting the size and growth of the housing stock. Secondly, the ageing trend of private housing is analyzed and we identify the scale of investment on managing and maintaining it. Then we discuss building-related accidents and recent survey results on the condition of these building. Lastly, we explore the severity of unauthorized building works in private housing.

Most domestic buildings in Hong Kong are high-rise buildings of framed reinforced concrete. Before the 1960s these buildings were normally three to six storeys high. With the advance of construction technology, domestic buildings of about 20 storeys have been constructed since the 1970s. Nowadays, high-rise domestic buildings of about 40 storeys have become the norm in the Hong Kong property market. A small object falling from this height, such as a debonded mosaic tile from the external wall, may result in fatal consequences. Not only is the initial quality of building critical, but good maintenance and management of these buildings is also indispensable.

2   Number of Housing Units

Table 2.1 shows the breakdown of the domestic building units. The domestic portfolio, of about 2 million flat units in 2001, is almost equally shared by private and public housing.

Table 2.1   Breakdown of Domestic Building Units by Ownership

Domestic Use	Number of Flat Units	Percentage (%)	Estimated GFA (m²)
Private	1,025,913	49.64%	54,880,455
Public[1]	1,040,902	50.36%	52,288,180
Total	2,066,815	100.00%	107,168,635

Source: Government Information Services (2001).

We estimate that domestic building units amounted to 107 million square metres gross floor area (GFA) at the end of 2000 and dominate the total portfolio of built facilities[2] in Hong Kong. Figure 2.1 shows that domestic built facilities account for more than 70% of the portfolio, in terms of GFA.

Figure 2.1   Distribution of GFA-quantified Built Facilities

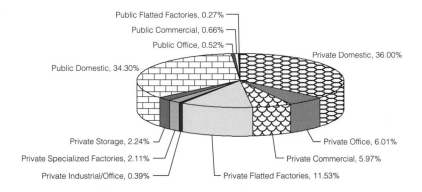

Figure 2.2 shows the growth in the number of private housing units in the past 50 years in Hong Kong. Except for the period during World War II, when a great deal (20%) of domestic accommodation was destroyed or seriously damaged, we have been facing a high speed of growth in terms of the number of private housing units. It has taken less than 10 years to increase 200,000 million flat units to the housing stock.

In 1955 the Buildings Ordinance was revised to permit significantly higher densities of development. This, together with a keen demand for housing led to a private sector development boom and the demolition of dilapidated premises climbed to a climax in the 1960s. Figure 2.3 depicts the growth rate of private housing units. It shows that during the 1960s, the growth rate reached as high as 12% per annum (p.a.), which can be considered as a period of building boom. More recently, the growth rate has been levelling off to about 3% p.a., which is about 30,000 new units per year.

Figure 2.2   Number of Private Housing Units

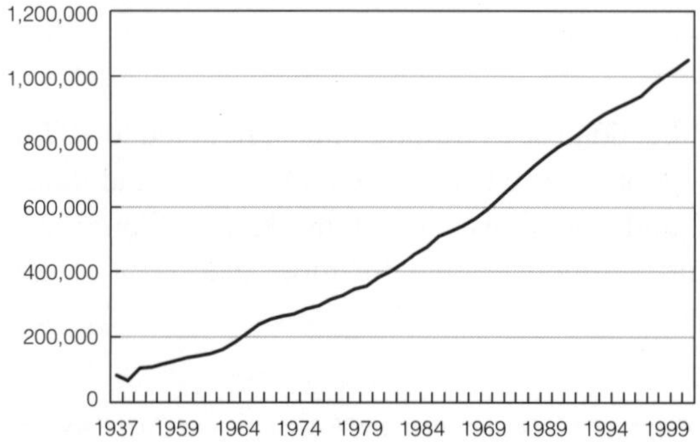

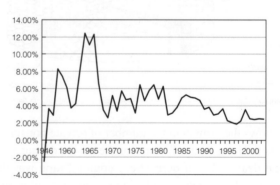

Source: Government Information Services (1936–2002), *Hong Kong Annual Report.*

Figure 2.3   Growth Rate of Private Housing Units

Source: Government Information Services (1936–2002), *Hong Kong Annual Report.*

3 Ageing Trend

During the 1960s we had a rather young portfolio of housing stock. There were only 21,000 pre-1903 premises in 1959, which constituted only 16% of the total stock of housing. The eradication of the dilapidation problem in the 1960s and 1970s could be attributed to the high rate of replacement. The number of pre-World War II premises was greatly reduced from 67,200 in 1946 to 30,000 in 1968 and then to 1,700 in 1991[3]. The proportion of old buildings dropped to an unprecedentedly low level in the 1970s. However, the replacement rate has become stagnant in recent decades.

The replacement rate is the proportion of the number of units demolished to the total housing stock one year before. It can be derived from the time series of housing completions and the number of the housing stock as in the formula of Malpezzi *et al.* (1987):

$$\Delta K_t = C_t - \delta K_{t-1} \tag{3.1}$$

where δ is the replacement rate, K_{t-1} is the housing stock in year *t-1*, C_t is the construction completion in year *t*, and $\Delta K_t = K_t - K_{t-1}$.

Barras (1983) estimated that the replacement rate in the UK ranged from 1% to 4% and Malpezzi *et al.* (1987) estimated it was from 0.6% to 1.3% in the USA. However, Figure 2.4 shows that the replacement rate of private housing in Hong Kong was on average 0.32% p.a. during 1980–2000, in comparison with the annual replacement rate of 2.6% in 1957–1966. It is noted that the quantity of the supply of new housing has been persisted at a high level, whereas the quantity of the demolition of housing stock has dropped to a negligible amount.

Figure 2.4   *Replacement Rate of Private Housing Units*

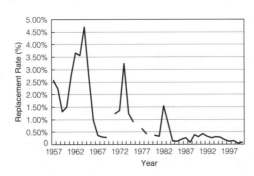

Source: Rating and Valuation Department (various issues).

The Housing, Planning and Lands Bureau (1998) estimated that in 2005 more than 40% of the private housing units in urban area will be over 30 years old and will therefore be reaching obsolescence. The sudden upsurge of the proportion of ageing houses is the result of the building boom in the 1960s. The stock distribution of different age groups vividly shows the dominance of this age group over the years (bars with a star in Figure 2.5). The building condition of this group of buildings has far-reaching implication for the building dilapidation problem in Hong Kong.

When housing units older than 30 years old are taken into consideration, their proportion to total housing stock has increased from 1% to 40% since 1976. A great wave of problems arising from this ageing housing stock in the coming future can be envisaged.

Figure 2.5   Distribution of the Building Age of Private Housing

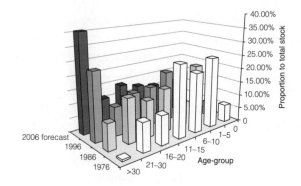

Source: Estimated from the data in the Rating and Valuation Department (1957–2002), Annual Report.

4 Investment on Building Maintenance and Management

Investment in the maintenance and management of buildings has long represented a relatively small part of the economy in Hong Kong (Ho, 1994, Baldwin, 1993), Figure 2.6 shows that the total gross value of maintenance work done (GVM) and the total service income of real estate maintenance management (REMM) represented only about 1.8% and 1.6% of the gross domestic product (GDP) respectively in 2001. It also reveals that GVM accounted for just less than 12% of total gross value of construction work (GVC) done in 2001. Investment on Building Maintenance and Management.

Investment in the maintenance and management of buildings has long represented a relatively small part of the economy in Hong Kong (Ho, 1994, Baldwin, 1993), Figure 2.6 shows that the total gross value of maintenance work done (GVM) and the total service income of real estate maintenance management (REMM) represented only about 1.8% and 1.6% of the gross domestic product (GDP) respectively in 2001. It also reveals that GVM accounted for just less than 12% of total gross value of construction work (GVC) done in 2001.

Figure 2.6   Economic Importance of Building Maintenance and Management

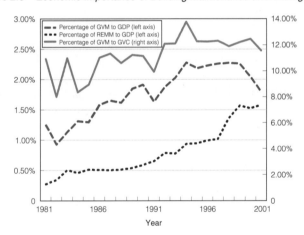

Source: Census and Statistics (various issues), Report on Annual Survey of Building, Construction and Real Estate Sectors, Hong Kong.

However, the rapid increase of GVM and REMM is supported by the following facts, as shown in Figure 2.7.

- the gross value of work done of maintenance work (GVM) in real terms has increased by 3-fold in the past two decades;
- the service income of real estate maintenance management (REMM) in real terms has increased by 13 times in the past two decades.

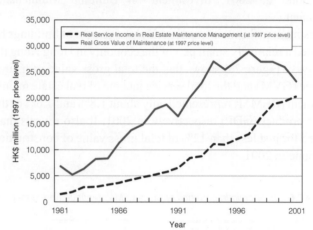

Figure 2.7   Real Gross Value of Building Maintenance and
Real Service Income of Maintenance Management

Source: Census and Statistics (various issues), Report on Annual Survey of Building, Construction and
Real Estate Sectors, Hong Kong.

The above figures clearly demonstrate that income from REMM has grown continuously, but GVM has been decreasing since the Asian financial turmoil in 1997. Taking the ageing trend and the growing number of building into consideration, a decrease of GVM will have an undesirable effect on the upkeep of the existing buildings.

5   Building-related Accidents of Private Housing

Every year there are some building-related accidents with casualties, especially those involving collapses, falling objects and fires. Table 2.2 shows the number of building-related accidents in private housing since 1990. It reveals an increasing trend which is probably the consequence of the ageing problem.

Table 2.2   Accidents Related to Private Building Since 1990

Year	Collapses	Falling Objects	Fires	Others	Total
1990	4	0	0	0	4
1991	0	0	0	0	0
1992	3	0	0	0	3
1993	4	1	0	0	5
1994	4	2	1	1	8
1995	3	0	2	0	5
1996	9	6	1	0	16
1997	9	2	2	0	13
1998	5	0	5	0	10
1999	5	16	6	0	27
2000	14	14	5	0	33
2001 (up to 18/04/01)	11	7	1	0	19

Source: Task Force on Building Safety and Preventive Maintenance (2001), Accidents Related to Building Safety since 1990, Home page of the Task Force, Hong Kong SAR Government.

6   Building Conditions

6.1   Pre-World War II Premises

The number of pre-World War II premises was reduced greatly to less than 1,700 in the 1990s (Chan, 1991). Most are located in Hong Kong Island and are under the close surveillance of the Buildings Department. Apart from those statutorily designated monuments and heritage buildings, the majority are in a poor condition and have been repaired to a barely marginal standard. Their low standard of design and construction is one of the reasons for their becoming dilapidation beyond repair.

> Many of them were constructed with timber joists and load-bearing brick wall. The mortar used for bonding the brickwork was normally composed of yellow earth and sand (and lime). These types of mortar are not expected to last long.   (Buildings Department, 1993)

6.2   Post-World War II Premises

The latest cycle of inspection on post-World War II premises can be traced back to the late 1980s when the general building condition had deteriorated to a crisis level. A large sample of private buildings (22,500 blocks, which accounted for about 41% of total stock) was inspected and the report was completed in early 1990. Blelengberg (1990) reported that 8,000 (35%) buildings were in a poor state of repair, 7 blocks have had to be closed and 67 blocks were in a dangerous state. The remaining 14,000 buildings would not require major repairs within 5 years.

In the wake of this alarming result, a complete survey of the whole population of private housing stock was carried out. In late 1990, the Building Authority acknowledged that almost 17,000 (31%) blocks of housing in Hong Kong were in a suspect structural condition (out of the 55,000 private housing blocks in total), while another 193 (0.35%) were in a dangerous condition, requiring urgent attention (Stoner 1990, Hughes, 1991). The buildings were sub-divided, according to the remedial action required, into three categories, as shown in Table 2.3.

Table 2.3   The Condition of Private Domestic Buildings in Hong Kong, 1992

Category	No. of Buildings	Conditions
Top Priority	210 (0.35%)	In immediate danger of collapse, requiring remedial action.
Category II	17,000 (31%)	Suspected to be dangerous, requiring detailed structural examination.
Low Priority	38,000 (69%)	Do not require major remedial action within the following 5 years.

Source: Leung (1992).

The Buildings Department completed the survey of the 16,700, category II private buildings in 1996 and 6,187 (37%) statutory orders were served, requiring certain buildings to be repaired or demolished. However the condition of building dilapidation did not seem to have been improved. In the wake of the collapse of a balcony at Bowring Street in 1993, a consultant was appointed by the Building Authority to conduct a condition survey of 4,308 privately owned post-World War II (built in 1946–1958) buildings with cantilevered elements. The Buildings Department's (1995) report was completed in 1995 and the findings raised concern on the durability of the buildings in this age group.

6.3   Premises built in 1946–1958

Firstly, the buildings of this group are inherently inferior because of the less stringent statutory requirements and technical know-how at that time.[4] Worse still, they were regarded as poorer even than those built in the 1940s because of the inferior workmanship associated with this pace of development. For example, the consultant who carried out detailed investigations on four buildings in this cohort found that:

reinforcement corrosion was actively occurring and was a
significant concern. At most locations examined, concrete
was of poor quality and in low compressive strength.[5] The
steel reinforcement concrete was carbonated and chloride
contaminated.[6] Section loss on stirrups often exceeded
25% and in some cases exceeded 50%. In almost all cases
the depth of carbonation exceeded the depth of concrete
cover to the steel by a considerable margin implying that
the steel may have been actively corroding for more than 30
years in some cases.   (Buildings Department, 1995)

Before any measures were taken, another balcony at Marble Road, which
fell in the 1946–1958 category, collapsed in 1997. The Buildings
Department (1998b) found that "it was caused by, *inter alia*, over-stressing
of the steel bars due to additional loading and misplacement of bars;
corrosion of steel bars together with concrete strength 30% below the
original design requirement". A special task force was set up to
investigate the structural condition of 645 reinforced concrete
cantilevered balcony structures in this age group. The results were
publicized in 1998 and are shown in Table 2.4.

Table 2.4   Balconies Conditions in Private Buildings, 1998

Type	Number	Conditions
Balconies built in 1946–1958	114 (18%)	Reinforcement showed rusting and damage in different degrees, which require detailed structural examination.
	531 (82%)	Do not require major remedial action within 5 years.

Source: Buildings Department's press conference reported by *Ming Pao Daily* (12 December 1998).

Premises built during 1959–1980

It is a common belief that dilapidation is merely the consequence of
deterioration, which is age-specific. The poor condition of buildings in
the age group 1946–1958 is simply because they are old and the problem
will soon be eradicated through redevelopment. However, survey results
refute this assumption. Another cohort of buildings built between 1959
and 1980 was also found to be very seriously dilapidated. This is believed
to be the consequence of the poor quality of construction materials and
workmanship during this period. For example, it is well known that during
1964–1965, the severe weather resulted in a harsh construction
environment. The Government recorded this as follows:

> In 1964, … the year has been notable for a very long dry
> spell of weather, which produced a draught of exceptional
> severity. Work requiring the use of clean water, such as
> concreting, plastering and *rendering*, has been held up.
>   (Government Information Services, 1964)

> In 1965, a record number of typhoons and tropical storms
> and shortages of granite aggregate was encountered.
>   (Government Information Services, 1965)

Unfortunately, the quantity of buildings supplied reached an
unprecedented peak in these years because of, *inter alia*, the amendment
of the Buildings Ordinance. In light of the provisions of the Building
(Planning) Regulation (Amendment No. 2) in 1962, the intensity of
development was to be more restricted.[7] It triggered a rush of
development plan submissions from private developers, as recorded in the
Hong Kong Annual Report, 1964:

> Private developers initiated a phenomenal number of
> schemes with intensities of development which new
> legislation was about to restrict: as a result the value of
> completed buildings in 1964 reached the unprecedented
> figure of $838.4 million in spite of unusually bad summer
> weather.   (Government Information Services, 1964)

There were two serious collapses of a canopy from buildings of this cohort
in 1994 and 1997. The Buildings Department (1994) reports on the collapse
of a canopy at Aberdeen in 1994. This revealed that, besides the
unauthorized use and alteration of the canopy, the position and spacing of
reinforcement was one of the causes of the accident. The building was
completed in 1973. This incident raised concern on the safety of the
canopies and the construction quality of buildings in this cohort. The
Authority conducted a systematic investigation programme on 981
reinforced concrete, cantilevered slab canopies. The results showed that
more than 10% of these canopies were dangerous and about 50% were
found to have a different extent of rusting of reinforcing bars, incorrect
position of reinforcement, overloading and/or over thick of concrete
(Table 2.5).

Table 2.5   Conditions of Canopies in Private Buildings, 1998

Type	Number	Condition
Cantilevered Canopy	100 (10%)	I immediate danger of collapse, requiring remedial action.
	476 (49%)	Suspected to be dangerous: requiring detailed structural examination.
	405 (41%)	Do not require major remedial action within 5 years.

Source: Buildings Department's press conference reported by *Ming Pao Daily* (12 December 1998).

Then another canopy at Kwun Tong collapsed in 1997, a building that had also been completed in 1972. The Buildings Department (1998a) report showed that the collapse of the canopy was the result of a failure in bending which was caused by, *inter alia*, the incorrect position of steel reinforcing bars during construction. The misplacement of steel bars was the common cause of both collapses.

A consultant was appointed to carry out a condition survey of buildings constructed between 1959 and 1980 and the cantilevered canopies in this age group to establish their deterioration trend. A total of 12,364 buildings were inspected. Detailed investigations on 1,004 buildings were conducted on selected structural elements and 497 cantilevered structures were assessed, leading to the publication of the Buildings Department (1999) report. Figure 2.8 shows the condition grading for different cohorts in their study and reveals that the cohort of 1960s shows the poorest building condition.

Figure 2.8   Inspectors' Condition Code on 1959–1980 Cohort of Buildings

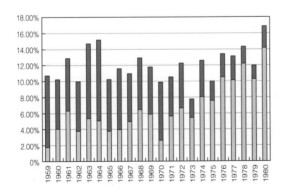

Source: Buildings Department (1999).

Table 2.6 shows the carbonation depth, chloride content and cement content of different building cohorts. These are considered to be the factors responsible for the deterioration of reinforced concrete. The performance of the 1959–1965 cohort was particularly poor. In terms of chloride content, 9% of the 1959–1965 cohort was found to contain chlorides above the corrosion threshold (0.4%). The 1970–1975 cohort was the next highest group found to contain calcium chloride above the permitted threshold. The report concluded that this was the result of the periods of fresh water shortages.

Table 2.6   Carbonation, Chloride Content and Cement Content of
Private Buildings, 1999

Cohort	Carbonation Depth (mm)		Chloride Chloride Content (% weight of sample)		Cement Content (%)	
	Range	Mean	Range	Mean	Range	Mean
1959–1965	5–135	50	0.01–1.58	0.04	2.7–41.8	13.6
1966–1970	0.5–120	41	0.01–0.52	0.03	6.3–30.8	14.2
1971–1975	0.5–105	35	0.01–0.57	0.04	6.3–33.1	18.9
1976–1980	0.5–101	27	0.01–0.27	0.02	10.6–34.4	17.8

Source: Buildings Department (1999).

The poor quality of construction materials in the 1960s is also reflected in the survey results of public rental housing estates in 1986. The results showed that 49.6% of the core strength were below specifications. Most of the below-specification cores came from housing built before 1974 (Table 2.7). The Housing Department (1986a, 1986b) reports concluded that 26 blocks of this housing were beyond repair and had to be demolished. Furthermore, a resolution was passed in 1987 by the Hong Kong Housing Authority that all public rental housing constructed before 1972 had to be demolished (Bates, 1996).

Table 2.7   The Results of Core Strength in Public Rental Housing, 1986

Cohort	Percentage of Core Sample with Different Strength (Specified Strength = 20 MPa)		
Strenght (MPa)	< 10	10–20	> 20
Pre 1960	.0	46.5	53.5
1960–1964	.0	49.4	50.6
1965–1969	.5	69.7	29.8
1970–1974	11.6	69.4	19.0
1975–1979	.0	8.7	91.3
1980–1982	.0	2.0	98.0
Post 1982	.0	0.0	100.0

Source: Housing Department (1986b).

6.4   Effectiveness of Repair

It is commonly conceived that the dilapidation of buildings is attributable solely to the lack of maintenance and repair. Investigation or repair orders are normally served[8] to landlords owning the defective elements. However, the Buildings Department (1995, 1999) reports raised concerns on about the reparability or the effectiveness of repair on buildings with a low level of durability. It noted that up to 20% of the buildings which had been repaired during the last 5 years still showed significant damage in various elements. This was imputed to inadequate standards of repair, inappropriate methods of repair or insufficient consideration of long-term durability.

The Marble Road canopy collapse case may shed light on the effectiveness of repair on buildings with poor durability. The Buildings Department (1998b) report pointed out that the dilapidation of the building had been identified since 1972 and four repair orders had been served on the owner in 1972, 1979, 1984 and 1992, mainly for reinforced concrete repairs. The former three had been complied with, while the last one was completed by the government contractor. This implies that the repair standards and methods of the last repair were in accordance with the Building Authority's requirements. Before the canopy collapsed in 1997, the building was inspected again in 1995 and an advisory letter was issued for the repair of loose external rendering and so on. The fact was that the building was actually intrinsically irreparable, as the report indicated that the steel bars were wrongly placed and the strength of concrete was inadequate even according to the standard of the original design. In view of the inexorable accelerating trend of depreciation of a property, landlords will simply neatly cover over an inherent construction quality problem in response to a repair order.

The Buildings Department (1999) report observed that many poor quality repairs or inappropriate methods had been applied which concealed the deterioration developing behind the repair.

6.5   Unauthorized Building Works

It was estimated that there were 800,000 illegal structures or unauthorized building works (UBWs) in 2001, and there have been about 10,000 new UBWs constructed every year (*Ming Pao Daily* 2000a). The following are examples of common UBWs in Hong Kong:

1. Cages, canopies, metal flower racks and any projection from the external wall of a building;
2. Canopies and structures that project over government land, pavements, or lanes;
3. Structures on rooftops, flat roofs, yards, or light wells and
4. Metal supporting frames for air-conditioning plants and cooling towers.

Figure 2.9 shows two typical examples of UBWs. The left photo shows a cage built on the canopy and the right photo shows a balcony projection from the external wall of a building.

Figure 2.9   Typical Examples of UBWs

Illegal Structure

Illegal Balcony

The danger of having numerous UBWs in a highly densely built city like Hong Kong is easy to see. Table 2.8 shows a non-exhaustive sample of the accidents that have been traced to UBWs since 1990. They have resulted in at least 21 deaths and 135 injuries from 1990 to 2002.

Table 2.8   Sample of Accidents Related to UBWs, 1990–2002

Date	Accidents	No. of Deaths	No. of Injuries
17 Aug 1990	Collapse of a canopy with UBWs in Mong Kok	1	0
27 Oct 1990	Collapse of a canopy with UBWs in To Kwa Wan	6	9
15 Oct 1993	Collapse of a balcony with UBWs in Yau Ma Tei	0	4
01 Aug 1994	Collapse of a canopy with UBWs in Aberdeen	1	16
15 Nov 1995	Collapse of an illegal canopy in Kwun Tong	1	2
16 Apr 1997	Collapse of a canopy in Kwun Tong	1	0
19 Jul 1997	Collapse of a balcony in North Point	0	5
21 Oct 1997	Collapse of an illegal cantilevered metal cage in Mong Kok	0	1
06 Jan 1998	Fire in unauthorized alterations in an exit route in North Point	2	49
31 Jul 1998	Collapse of an illegal canopy in Kwun Tong	1	3
14 Sep 1998	Fire in tin huts illegally built in Wan Chai	2	13
03 Oct 1998	Fire on a rooftop with illegal structures in Mong Kok	0	0
11 Dec 1998	Fire on a rooftop with illegal structures in Sau Mau Ping	0	0
17 Jan 1999	Fire on a rooftop with illegal structures in North Point	0	0
09 Feb 1999	Fire on a podium with illegal structures in Kwai Chung	0	5
24 Feb 1999	Fire in a flat with an illegal alteration in Mong Kok	0	0
07 May 1999	Fire in an illegally built workshop in Kwun Tong	0	0
10 Aug 1999	Falling of masonry from an illegally built canopy in Mong Kok	1	0
11 Aug 1999	Collapse of an illegally built ceiling in North Point	0	0
10 Sep 1999	Collapse of an illegally built ceiling in Mong Kok	0	1
03 Oct 1999	Collapse of an illegally built podium in Tai Kok Tsui	0	2
22 Nov 1999	Fire in illegal structures behind a building in Yau Ma Tei	1	8
01 Dec 1999	Fire in an illegal structure on a rooftop in Sham Shui Po	1	2
02 Mar 2000	Fire in an illegal rooftop structures in Tsuen Wan	2	5
02 Dec 2000	Fire in an illegal rooftop structure in Hung Hom	0	0
02 Mar 2001	Fire in an illegal rooftop structure in San Po Kong	0	0
07 Mar 2001	Collapse of external walls of illegal rooftop structures during demolition in Ngau Tau Kok	0	0
17 Apr 2001	Collapse of an illegally built canopy in Kowloon City	0	1
08 Jun 2001	Collapse of the roof of an illegally built unit in Chai Wan	0	0
25 Mar 2002	Collapse of an illegal balcony in To Kwan Wan	0	7
11 Aug 2002	Collapse of an illegal balcony during demolition in Kwun Tong	1	2

Source: Lai and Ho (2001); Task Force of Building Safety and Preventive Maintenance (2001).

Throughout the last three decades removing UBWs has been like an elephant trying to get rid of a mouse running over its body. Different district boards co-operated with the Building Authority to carry out the removals, but they achieved only short-term results and the UBWs were soon rebuilt. There were also several large-scale operations to remove UBWs. For instance, in 1983 the Shum Shui Po District Board set aside $150,000 to remove 1,000 illegal structures. During the first phase, 170 structures, 57 canopies and 54 shop extensions were removed. The second phase saw the removal of 481 illegal structures (South China Morning Post, 1984a). In 1984 the government launched one of its biggest clearance operations against a multi-storey building in San Po Kong with nearly 650 illegal extensions (Chan, 1984). In 1991, the Operation Appendages, initialized by the Building Authority, targeted illegal structures on 30 buildings: as a result, 3,800 statutory orders were served, demanding the removal of 5,800 illegal additions. Another scheme, the Operation Catherine Wheel, carried out by the Building Authority, focused on 52 buildings and issued 6,800 advisory letters about the removal of 11,160 illegal projections. These operations were deemed successful, with a 55% clearance rate (Sinclair, 1993). However, after the aforementioned series of incidents in 1999, the issue of UBWs again made headlines. The Building Authority set out to eradicate the problem, and served 17,000 advisory letters and 7,600 statutory orders on the owners of 300 blacklisted buildings that housed over 8,000 illegal structures. Even UBWs that had been built before 1975 were not spared, in view of the high risk of the targeted buildings (*Ming Pao Daily*, 1999).

The Buildings Department has conducted a series of large-scale blitz clearance operations since 2001. These operations mainly target the removal of UBWs on external walls. For example, on 7 May 2001 a large-scale blitz clearance operation was conducted in 13 streets of Mong Kok and Tim Sha Tsui, covering 330 aged buildings (Ming Pao Daily, 2001a). Another operation was held in June 2001. About 130 professional and technical staff members were mobilized to take part in a three-day operation that covered 85 buildings in Western District, 153 buildings in Tai Kok Tsui and 71 buildings in Tsuen Wan.

The Buildings Department (2001) reported that "with the stepping up of blitz operation last years, the Buildings Department has succeeded in removing more than 20,000 unauthorized building works, instead of the original target of 15,000."

Table 2.9 shows the number of reported UBWs and the number of UBWs that were removed from 1990 to 2002. It reveals a significant increase in the number of removals after 1999. The average number of removals per annum from 1990 to 1998 was only 4,853, while the average number of removals per annum from 1999 to 2001 increased greatly to 12,617 (a 160% increase). Table 2.10 shows also the dramatic increase in the number of buildings with UBWs that were targeted for clearance.

Table 2.9   No. of Reports Received and No. of Removals of UBWs

Year	Reports Received of UBWs	Removal of UBWs
1990	7,009	2,269
1991	7,420	6,857
1992	6,992	6,969
1993	8,437	8,759
1994	7,596	4,890
1995	8,203	3,883
1996	9,913	3,479
1997	12,427	3,103
1998	13,163	3,471
1999	17,014	14,038
2000	15,860	10,602
2001	13,817	13,212
2002 (target)	11,000	25,000

Source: Lai and Ho (2001); Buildings Department (2002); *Ming Pao Daily* (2000b).

Table 2.10   No. of Buildings Targeted for Clearance of UBWs

Year	1999	2000	2001	2002
No. of Buildings Targeted for Clearance of UBWs	100	404	1,574	1,500

Source: Buildings Department (2002).

However, this improvement is not achieved without a cost. The government invested more than $0.3 billion in recruiting additional staff members and enhancing the efficiency of removing UBWs in 2000–2002. An estimated 3,000 buildings were inspected during that time. In other words, each inspection cost approximately $5,000 in 1999–2002 and it will require another $4 billion just to inspect the existing UBWs (without accounting for the costs of removal). Worse still, it will take another 63 years to remove completely the existing 800,000 UBWs in Hong Kong, assuming that no new UBWs will be erected and the existing removal rate is maintained.

7   Conclusion

We have experienced a very long period of private building dilapidation after World War II. Yet, to start with the problem was tackled by the massive replacement of units. Similarly, we have had the unpleasant experience of poor construction quality in public rental housing in the

1970s. This was also handled by large-scale redevelopment schemes. Although the problem of pre-World War II premises has more or less dwindled away and a lot of poor quality public rental housing has been pulled down, we have paid a heavy cost for this and we cannot afford to have similar problem again.

However, a new spate of building-related accidents seems to be approaching. This time it will create much stronger havoc because of the mode of high-rise and high-density developments since 1955. Much has been done to ameliorate the situation, but the results are not promising. On the contrary, an acceleration of the rate of deterioration is envisaged if the situation is tolerated. This is because (i) the rate of replacement has almost come to a halt; (ii) the huge housing stock completed in the 1960s boom is now 30 years old; (iii) the durability of the buildings of the 1950s and 1960s cohorts is inherently low; (iv) there is widespread ignorance of the maintenance and management of building and (v) a much great number of unauthorized building works exists.

Notes

1. It includes 650,314 units of public rental housing provided by the Hong Kong Housing Authority; 278,268 units sold under Home Ownership Scheme; 53,905 units sold under Tenant Purchase Scheme; 32,337 units of rental housing governed by the Housing Society and 26,079 units of government quarters.

2. Non-GFA quantifiable built facilities, such as hotels, public utilities, etc. are excluded.

3. Rating and Valuation Department (various issues).

4. Most of the buildings built between 1946 and 1958 were designed to either the 1915 or 1938 reviews of the London County Council By-Laws. Not only is the concrete strength required much lower, but the elements designed to previous codes also have inherently lower safety margins in shear than that in the current code. The report figured out that it was very common to find elements designed to these old codes with a lower capacity in shear.

5. The mean compressive strength approximated 11MPa with a standard deviation of 3MPa. Although it is far lower than would be expected today, the results were not far from the strength required by the original design standard (about 12.5MPa).

6. Chloride ion concentration was typically three times the corrosion threshold of 0.4% chloride ion by weight of cement.

7. Before the amendment, the intensity of development was controlled by means of the volume of a building calculated on street width and permissible wall heights. But under the provisions of the amendment, the intensity of development is controlled by the use of plot ratio and site coverage. The amendment was passed in 1962, but it will not be fully operative until Jan 1966. Certain relaxation of time limits was allowed later in the wake of building bust since 1965.

8. vested in s.26A and s.26 of the Buildings Ordinance (Cap. 123).

References

1. Baldwin, G. R. 1993. *Property Management: A Review of Current Practice and Trends in Hong Kong*. Department of Building and Construction, City University of Hong Kong, September 1993.

2. Barras, R. 1983. A Simple Theoretical Model of the Office-Development Cycle. *Environment and Planning A*, *15*, 1361–1394.

3. Bates, R. 1996. Self Regulation in Public Housing. *1996 World Organization of Building Officials (WOBO) Fourth World Congress*, 2–8 November 1996, Hong Kong.

4. Blelenberg, K. 1990. Owners Could Face Big Bills for Renovation. *South China Morning Post*, 10 January.

5. Buildings Department. 1993. *Report on the Collapse of the Buildings at Nos. 28–30 Wing On Street, Hong Kong*, Hong Kong.

6. _____. 1994. *Final Report on the Collapse of the 1/F Canopy Albert House, 12 Sai On Street and 20–28 Chengtu Road, Aberdeen*, Hong Kong.

7. _____. 1995. *Report on Condition Survey of Private Reinforced Concrete Building Structures*. By Taywood Engineering Ltd, Hong Kong.

8. _____. 1998a. *Investigation Report on the Collapse of the Canopy at 418A Kwun Tong Road (WKK Factory Building)*. November 1998, BD Report BD 1/98, Hong Kong.

9. _____. 1998b. *Report on the Collapse of 1/F Balcony at No. 1P Marble Road, North Point, Hong Kong*. November 1998, BD Report BD 2/98, Hong Kong.

10. _____. 1999. *Consultancy Study on Steel Corrosion and Material Deterioration of Buildings Completed between 1959 and 1980*. By Taywood-Arup-Vigers JV, Hong Kong.

11. _____. 2001. Contracts awarded to clear unauthorized building works, *Press Release* 25 December 2001.

12. _____. 2002. Building Department sets to achieve record-breaking targets. *Press Release*, 26 January 2002.

13. Census and Statistics Department HKSAR Government. 1981–2001. *Report on Annual Survey of Building Construction and Real Estate Sectors.* Hong Kong.

14. Chan, F. 1991. 18 Families Lose Homes in Dangerous Pre-War Building. *South China Morning Post*, 22 June.

15. Chan, J. 1984. Big Drive to Clear Illegal Extensions. *South China Morning Post*, 27 July 1984.

16. Government Information Services. 1936–2002. *Hong Kong Annual Report.* Hong Kong Government.

17. Ho, C. W. D. 1994. Maintenance Works Sectors in the Fast Track. *Discussion Paper Series* (unpublished). Department of Real Estate and Construction, University of Hong Kong.

18. Housing Department. 1986a. *A Report on the History Procedure and Findings of a Structural Survey of the Hong Kong Housing Authority Building Stocks.* January 1986, Hong Kong Government.

19. _____. 1986b. *Summary Report on Structural Investigation on Final Report Addendum to Interim Report Issued January 1986.* April 1986, Hong Kong Government.

20. Housing, Planning and Lands Bureau. 1998. Urban Renewal in Hong Kong. Homepage of the Bureau: *www.plb.gov.hk/renewal*, Hong Kong SAR Government.

21. Hughes, O. 1991. Owners of Faulty Buildings May Face Daily Fines. *South China Morning Post*, 15 March.

22. Lai, L. W. C. and Ho, D. C. W. 2001. Unauthorized Structures in a High-Rise High-Density Environment: The Case of Hong Kong. *Property Management*, 19, 2, 112–123.

23. Leung, J. 1987. Increased Checks on Defective Housing. *South China Morning Post*, 15 May.

24. _____. 1992. Experts knew of risk at buildings. *South China Morning Post*, 9 Oct.

25. Malpezzi, S. Ozanne, L. and Thibodeau, T. G. 1987. Microeconomic Estimates of Housing Depreciation. *Land Economics*, *63,4,* 372–385.

26. *Ming Pao Daily*. 1998. 400 Canopies are Potentially Dangerous. *Local News*, 12 December 1998. (Chinese)

27. _____. 1999. 300高危大廈曝光 (300 High risk buildings are identified). *Ming Pao Daily*, 22 October 1999 (Chinese).

28. _____. 2000a. 三年拆五萬外牆僭建物 (50,000 UBWs will be demolished in 3 years). *Ming Pao Daily*, 15 March 2000 (Chinese).

29. _____. 2000b. 鄰近修簷篷街坊繞道行 . *Ming Pao Daily*, 26 March 2000: A2.

30. _____. 2001a. 屋宇署旺角清拆僭建 (Buildings Department Cleared UBWs In Mongkok. *Ming Pao Daily*, 7 May 2001: A8.

31. _____. 2001b. 屋宇署派員拆 30 大廈僭建物 (Buildings Department dismantled 30 UBWs) . *Ming Pao Daily*, 15 June 2001: A4.

32. Rating and Valuation Department. 1957–2002. *Property Review*. Hong Kong SAR Government.

33. Sinclair, K. 1993. Campaign to Remove Illegal Structures Proves Successful. *South China Morning Post*, 10 February 1993.

34. *South China Morning Post*. 1984. Fresh Drive at Illegal Building. *South China Morning Post,* 4 Feb 1984.

35. Stoner, T. 1990. Four More Buildings in Poor State. *South China Morning Post*, 14 October.

36. Task Force on Building Safety and Preventive Maintenance. 2001. Accidents Related to Building Safety since 1990. Home page of the Task Force: *www.plb.gov.hk/taskforce*, Hong Kong SAR Government.

3

Urban Renewal in Hong Kong — Historical Development and Current Issues

P. K. Kam, Sik Hung Ng and Charles C. K. Ho

The Urban Renewal Authority (URA) has done a lot to help improve the living conditions of many residents in old urban areas. On paper, its work is well-intentioned, comprehensive and beneficial to the Hong Kong community; but in practice thus far there are problems. In this chapter, we have identified some of the problems from the point of view of affected residents. Our stress on problems rather than on achievements in no way denies or undervalues the latter, but is meant to be constructive by giving voice to residents who, for various reasons, feel aggrieved. The chapter also discusses the implementation of the people-centred mission adopted by the URA.

Chapter

3

Urban Renewal in Hong Kong — Historical Development and Current Issues

P. K. Kam, Sik Hung Ng and Charles C. K. Ho

1 Introduction

In 2001, the Hong Kong Government established the Urban Renewal Authority (URA) to take a more proactive approach to improving the standard of housing and the built environment by undertaking, encouraging, promoting and facilitating urban renewal. Under the new Urban Renewal Authority Ordinance, the URA was granted a greater land resumption power to speed up urban redevelopment. In 2002 the URA launched three early-batch projects. To date, a total of 10 projects have been announced, all of which are carry-overs from the unfinished projects of the URA's predecessor, the Land Development Corporation (LDC). Apart from receiving greater land resumption powers and adopting a more proactive approach than the LDC, the URA also adopts a "people-centred" mission with respect to discharging its massive obligation of property acquisition and resident resettlement. It is now time to review the historical background and the development of urban renewal strategies in Hong Kong, the current work of the URA and the possible impacts of urban renewal. The purpose of this article is to discuss these issues.

The discussion in this paper is based on an interdisciplinary research project, "Ageing Building and People", collaboratively conducted by research teams from the Department of Applied Social Studies and the

Department of Building and Construction, City University of Hong Kong. One of the objectives of the research is to understand the impact of urban renewal on property owners and tenants in the redeveloped areas. As part of the research, a household survey was conducted to understand the situation of affected residents in urban renewal areas that were officially defined as environmentally poor and unsafe. Further information of the survey and some of survey results pertaining to "quality of life" can be found in chapter 6 of this volume. Focus groups were formed and in-depth interviews were also conducted in order to understand the different attitudes of residents towards urban renewal and the URA's mission, its compensation policy and its relationship with residents.

2   Historical Background of Urban Renewal in Hong Kong

Urban renewal in Hong Kong has a long history, and can be traced to slum clearance as early as 1884. However, active government involvement in urban renewal did not occur until 1960. In the period from the 1960s to the 1980s, the Government's urban renewal activities were mainly confined to improving traffic circulation, environmental conditions and community facilities in older and poorer areas. However, it did not give urban renewal top priority when dealing with the housing shortage. Instead, it took great efforts to develop "new towns" for housing large numbers of the urban population and for facilitating industrial development.

Only private developers took sporadic action to redevelop some of the older urban areas. Since the relaxation of the Buildings Ordinance in 1956 to allow the development of high-rise buildings on land previously occupied by low-rise buildings, the increase in land prices generated intense redevelopment activities by the private sector. Because of multiple ownership of the same building, it often took developers many years to assemble redevelopment sites by their own efforts. Thus, site assemblage represented a high cost for them in terms of time and compensation.

Owing to the immense difficulty of property acquisition, private-sector redevelopment was mostly in small-scale projects. This led to the construction of many "toothpick-like" buildings in urban areas which become an eyesore to visitors and local people. More seriously, small-scale redevelopment aggravated rather than relieved the intense congestion of the older urban areas because it focused on residential development but ignored the development of public open space or community facilities (Adam, 2000). As a result, it left the future comprehensive redevelopment of the surrounding building blocks in a difficult situation.

During the 1980s, in both Britain and the United States, developmental partnerships between the public and private sectors became a prominent feature of urban policy (Adams, 2000). The governments of these two countries believed that partnership with the private sector was an important resource for urban regeneration. Inspired by their redevelopment experiences, the then colonial Hong Kong government established the Land Development Corporation (LDC) in 1987 to undertake, encourage, promote and facilitate renewal within the older urban areas of Hong Kong. The LDC was given only HK$100 million by the government as a repayable loan. The government did not finance the LDC on a recurrent basis. In each renewal project, the LDC was required to pay a land premium and tax. To sustain urban renewal, the LDC had to select profitable renewal sites and follow strictly prudent commercial principles in its decisions and operations. A major mode of redevelopment in the LDC period was to attract private-sector resources to facilitate urban renewal.

The LDC was responsible for purchasing and resuming land and for planning and developing identified sites. However, it had no direct resumption powers and no access to public housing on behalf of the dislocated residents (Koo, 2001). By mid-1996, the LDC had completed 10 projects, with another 15 under development or active consideration. It relied on private negotiation to acquire properties. Since the LDC did not have land resumption powers and some residential unit owners demanded excessive prices from it, the pace of redevelopment was very slow. Besides, owing to the problem of multiple ownership, the LDC found it difficult to acquire ownership of all properties in a short period of time. Different owners with different interests and needs would demand different packages of compensation from the LDC. Complex issues such as these tended to prolong negotiation and thereby increase the cost of interest to the LDC.

Later, the LDC designed an owner participation scheme to involve owners in redevelopment. However the scheme failed to attract residential owners because of high development costs. So the difficulty of property acquisition remained unresolved right up to the establishment of the Urban Renewal Authority.

In 1997, the Asian financial crisis was a turning point for the LDC. The property market was cooling down. Many redevelopment projects did not look profitable to private developers due to the low plot ratio gain and the high acquisition cost. Under this "stagnant" property market situation, redevelopment projects failed to interest private developers. In 1998, LDC declared 25 sites as urban renewal projects involving HK$800 billion, but few private developers expressed any interest in them. As a result, these 25 redevelopment projects came to a halt.

In summary, after years of experiences, the problems of urban renewal under the auspices of the LDC include the following:

Site Assembly

Multiple ownerships make it difficult to assemble individual properties into lots for large-scale comprehensive redevelopment, as numerous separate legal interests have to be acquired first. This is difficult and sometimes impossible.

Relocation

Urban redevelopment necessitates the relocation of residents and businesses, resulting in high financial and social costs. The amended Landlord and Tenant (Consolidation) Ordinance, 1996, has increased the compensation rate for dispossessed domestic tenants.

Viability

There is a belief that redevelopment is highly profitable. This may have been the case, but is no longer necessarily so. The value of redeveloped properties may not cover redevelopment costs, particularly when there is little increase or even reduction in housing density after redevelopment.

3   New Urban Renewal Strategy

The Chief Executive of the Hong Kong Special Administrative Region Government in his 1999 policy address noted the urgent need for a new and proactive urban renewal strategy. He argued that any new approach should result in real improvements to the quality of life for people housed in dilapidated buildings in old and run-down areas, an important issue that will be addressed in Chapter 6 by Ng *et al*. in this volume. To further this aim, the Chief Executive announced in the policy address the establishment of a new Urban Renewal Authority to replace the LDC. This was followed by the development of a new Urban Renewal Strategy (URS), prepared by the Planning, Environment and Lands Bureau (PELB). A Report on the Urban Renewal Strategy Study was released in 1999. It was suggested in this Study that the new URS should aim at:

> Facilitating comprehensive planning over larger areas, providing additional green areas of open space, and community facilities and improving road networks while preserving the distinctive features of the old districts concerned.   (Planning Department, 1999, p. 1)

The Study also recommended that the future URS should adopt a two-pronged approach, namely, redevelopment and rehabilitation, to handle the problems of urban renewal.

The results of the Study provide a basis for the future direction of urban renewal. It has become an integral part of the overall strategic planning process in reshaping the old urban areas of Hong Kong. The Study clearly sets out the focus of future urban renewal which consists of redeveloping old and dilapidated areas, revitalizing stagnant old areas, rehabilitating run-down buildings and preserving buildings or places of local architectural, cultural or historical value. Instead of piecemeal redevelopment, the Study recommended that a comprehensive redevelopment area should be over $200m^2$ to achieve a better plot ratio and floor space efficiency. The new mode of urban renewal is to consider broader issues, other than building conditions and fire safety conditions. It can incorporate other planning needs such as the provision of sufficient government, institute and community (GIC) facilities, improvement of building and street layout, and the easing of traffic congestion and environmental degradation.

It was estimated by the Planning and Lands Bureau in 2001 that there were about 9,300 private buildings that are 30 years old and above, and that in ten years' time, the number of buildings over 30 years old will increase by 50% (Planning and Lands Bureau, 2001). The Planning and Lands Bureau also identified 200 new priority areas for urban renewal and an additional 25 uncompleted projects of the LDC. In physical terms 9,500 buildings occupying 55 ha. of residential land are on the priority list. In human terms, these buildings are home to more than 915,000 residents. The financial cost is estimated to be more than $HK250 billion.

Among the 225 projects, top priority is given to the 25 uncompleted projects of the LDC.

4   Urban Renewal Authority

4.1   Its Establishment and Powers

Following the Chief Executive's announcement in the 1999 policy address setting up a new Urban Renewal Authority, a Urban Renewal Authority Ordinance (Chapter 563) was drafted and introduced into the Legislative Council in February 2000. The Ordinance sets out the powers, duties, terms of reference, structure and composition of the URA. The proposed URA was empowered by the Ordinance to apply for land resumption without going through a protracted land acquisition process. Though there were heated debates and contradictory views about the merits of granting land resumption powers to the new URA as well as about the fairness of the compensation packages for the property owners during the process of public consultation, the Ordinance was eventually passed by the Legislative Council in June 2000. After the official endorsement of the Finance Committee of the Legislative Council on the compensation proposal for land resumption in March 2001, the URA was formally established on 1 May 2001.

In June, 2002, the Finance Committee of the Legislative Council approved a capital commitment of HK$10 billion to the URA over the next five years to enable the implementation of its urban renewal programme (Urban Renewal Authority, 2002a). With more financial and legal support from the government, the government expects the URA to be able to speed up urban redevelopment, facilitate urban rehabilitation and encourage the preservation of historical buildings.

The policy of the URA is guided by the finalized URS (Planning and Lands Bureau, 2001) which was published in November 2001. Under the framework of the URS, the URA is required to adopt a comprehensive and holistic approach to the rejuvenation of the older urban areas by way of redevelopment, rehabilitation and heritage preservation. It is given the task of a 20-year redevelopment programme involving 200 new projects and the 25 top-priority LDC incomplete projects in the nine target areas, which will affect some 126,000 people and 2,000 old buildings (Urban Renewal Authority, 2001).

To ensure that the renewal project is viable and meets social needs, the URA is required under the ordinance to prepare a five-year corporate plan and an annual business plan for approval by the Financial Secretary. In these plans, the number of new projects, the financial sources of redevelopment and the means of accommodating the dislocated residents must be clearly laid out.

Under Section 24 of the new Ordinance, the URA is also granted the authority to directly appeal to the Chief Executive for land resumption. With full government backing, the URA could assemble lands more quickly. Though affected owners may decline the URA's offer, the URA could make use of its land resumption power to acquire private properties in advance of a court of appeal hearing. Those who decline the URA's offer could appeal to the Lands Tribunal for a final decision on the compensation.

The URA is allowed to adopt three different ways to implement urban renewal projects. Firstly, it can redevelop resumed sites on its own. Secondly, it can collaborate with private developers. Thirdly, it can resume identified sites and put them out to tender during a better market environment. To leverage private sector resources, the government gives the URA financial and non-financial tools to attract the private sector. Financial tools include waiving the land premium. Non-financial tools include the exemption of government/institute/community facilities (GIC) from gross floor area (GFA) calculation and the relaxation of plot ratios to the current building ordinance maximum permitted level.

4.2　The Four Rs and the People-centred Mission

One of the features that differentiates the URA from the LDC is its adoption of a graded approach to urban renewal wherein redevelopment, which involves demolishing old and dilapidated buildings, is no longer the only option (as was the case of the LDC) but coexists with three others. The four options are commonly known as the "Four Rs": redevelopment, preservation, rehabilitation and revitalization (see http://www.ura.org.hk).

Redevelopment

This option focuses on demolishing old and dilapidated buildings providing poor living conditions and improving living conditions by assembling large pieces of land for comprehensive planning and the restructuring of the community through providing better and appropriate community facilities and open spaces.

Preservation

Unlike the LDC, the duties of the URA are not confined to demolishing old buildings. It is also responsible for preserving heritage buildings and protecting and conserving sites and structures of historical, cultural or architectural interest or value while developing old districts.

Rehabilitation

Rehabilitation is a new mandate for the URA from Government. It will play a major role in the new era of urban renewal because it will extend the useful life of buildings to reduce the need or urgency of redevelopment. The URA is thus asked to take up the duty of working with owners, the government and other parties to prevent or slow down the decay of the built environment by promoting proper maintenance of buildings in old urban areas.

Revitalization

The broader concern of new urban renewal is to revive and strengthen the economic and environmental fabric of old local districts. It involves the adoption of a more holistic, co-ordinated approach involving its partners and stakeholders to improve the quality of urban life through redevelopment, rehabilitation and preservation initiatives.

The Four Rs approach sets out the new priority and focus of urban renewal for the URA. It has made the work of the URA broader and more comprehensive than that of the LDC. But at the same time, it requires the input of more resources and more co-ordinated efforts among the URA and other government departments and concerned sectors.

Apart from the Four Rs approach, the URA also claims to adopt a 'people-centred' approach to implement its policies. As stated in the Urban Renewal Strategy, based on the people-centred approach, the purpose of urban renewal

> … is to improve the quality of life of residents in the urban area. The Government has to balance the interests and needs of all sectors of the community without sacrificing

the lawful rights of any particular group. The aim is to reduce the number of inadequately housed people.

(Planning and Lands Bureau, 2001, p. 1)

The principles underlying this approach include (see Fisher, 2001; Planning and Lands Bureau, 2001; Urban Renewal Authority, 2001):

1) owners will receive fair compensation,
2) all displaced tenants will be properly rehoused,
3) adverse impacts will be minimized,
4) the community must benefit (through upgraded facilities) and
5) affected residents should be given an opportunity to express their views on the redevelopment projects.

From these principles the following initiatives were launched:

1) To establish a social service team in each renewal district to provide counselling and advice to residents who have experienced problems or have concerns.
2) To set up the District Advisory Committee (DAC) to understand local urban renewal needs. The DAC comprises local residents, professionals, social workers, district council members, academics and the like. They mainly provide advice to the URA in relation to the planning and implementation of projects.
3) To carry out pioneering social impact assessments (SIA) for the redevelopment projects.
4) To set up an appeal panel to handle appeals by the affected residents on compensation and rehousing matters.

The people-centred approach is regarded as the guiding philosophy of the urban renewal work under the management of the URA. Judging from its mission statements, it seems that the URA tries to differentiate itself from the negatively perceived, profit-driven image of the past LDC and to promote a new image for itself as a people-centred entity. However, in our contact with affected property owners and tenants in the research, criticisms were heard that the URA's avowed people-centred approach was in practice more of a rhetoric than a real commitment. These criticisms and comments, to be discussed in a later section, can best be seen in the context of the current working of the URA.

4.3   Current Working of the URA

After its formation in May, 2001, the URA began its first year of operation. A major problem confronting the URA Board was to meet the Government's expectation that the urban renewal programme should be self-financing in the long run. Though the Government has approved the waiver of land premium for the URA and has injected a capital of $10 billion, considerable worry still exists over the financial viability of the urban renewal projects. Under the URA Ordinance, the URA has to take over all the assets and liabilities of the LDC. It is envisaged that there are potential losses from taking over some of the incomplete projects of the LDC. Therefore, right from the start, the URA is not in a strong financial situation and it needs to consider ways of raising revenues through joint venture partnerships with private developers and borrowing loans from banks.

The URA began its work by focusing on the 25 top-priority incomplete projects announced in early 1998 by the LDC. The work was thus concentrated more on redevelopment projects and the work on rehabilitation, preservation and revitalization was put to one side. The first Five-Year Corporate Plan and the Annual Business Plan were finally submitted to the Government and were approved in March 2002. The first business plan includes eight projects in five districts. All of them are among the 25 LDC projects. The three 'early launch' redevelopment projects which started prior to the submission of the Five Year Corporate Plan and the Annual Business Plan — Fu Wing street project in Sham Shui Po, Cherry Street project in Tai Kok Tsui and Johnston Road Project in Wan Chai — were announced on 11 January 2002. On 5 July 2002, another redevelopment project at the corner of Po On street and Shun Ling Street in Sham Shui Po was announced after the government's approval of the 5-year Corporate and Business Plan. Another two projects were announced in the year of 2002, namely the Reclamation Street Project in Mongkok (started on 18 October 2002) and the First Street/Second Street Project in Sai Ying Pun (started on 29 November 2002). In 2003, four more redevelopment projects were announced, including the Staunton Street/Wing Lee Street Project in Sheung Wan (started on 21 March 2003), the Queen's Road East Project in Wan Chai (started on 21 March 2003) and the Bedford Road/Larch Street Project in Tai Kok Tsui (started on 4 July 2003). In June 2003, the URA began its work on revitalization. The first revitalization project at Sheung Wan was announced on 20 June 2003. Table 3.1 summarizes the projects up to July, 2003.

Table 3.1   Summary of Urban Renewal Projects from January 2002 to July 2003

Date of Commincement	Project	Nature	Location	Site Area (m²)	No. of Buildings	No. of Households	No. of Affected Property Interest	Development Costs (HK$ million)
11 January 2002	Fuk Wing Street/Fuk Wa Street	Redevelopment	Sham Shui Po	1,362	8	559	74	380
11 January 2002	Johnston Road	Redevelopment	Wan Chai	2,062	20	206	104	880
11 January 2002	Cherry Street	Redevelopment	Tai Kok Tsui	4,327	33	137	238	1,300
5 July 2002	Po On Road/Shun Ling Road	Redevelopment	Sham Shui Po	1,380	8	177	72	400
18 October 2002	Reclamation Street	Redevelopment	Mong Kok	533	4	80	24	180
29 November 2002	First Street/Second Street	Redevelopment	Sai Ying Pun	3,511	30	458	293	1,140
21 March 2003	Queen's Road Street	Redevelopment	Wan Chai	380	6	40	18	160
21 March 2003	Staunton StreetWing Lee Street	Redevelopment	Sheung Wan	4,460	45	183	87	1,020
20 June 2003	Sheung Wan Fong	Revitalization	Sheung Wan					
4 July 2003	Bedford Road/Larch Street	Redevelopment	Tai Kok Tsui	1,236	7	309	76	433
4 July 2003	Barker Court Street	Redevelopment	Hung Hom	250	2	23	14	72
Total				18,265	163	2,172	913	5,975

Sources: Urban Renewal Authority (2003) and retrieved from http://www.ura.org.hk/html/c512000e1b.html

The process of land resumption and the acquisition of property has met with varying degrees of acceptance in different projects. By 7 July 2002 (about half a year after the announcement of the projects), the URA had succeeded in reaching agreements with about 80 per cent of the owners in the Johnston Road Project in Wan Chai, 72 per cent in Fuk Wing Street/Fuk Wah Street Project in Sham Shui Po, and 52 per cent in Cherry Street Project in Tai Kok Tsui. Overall, 60 per cent of households signed the agreements (Urban Renewal Authority, 2002b). The remaining 40 per cent of the affected owners rejected the URA's compensation offer.

Up to mid-2003, all tenants in the three "early launch" projects have been rehoused or received compensation. However, the acquisition of property takes time and is not easy as quite a lot of owners have objected to the compensation offer (*Sing Tao*, 19 April 2002; *Sing Pao*, 12 May 2002, *Sun*, 26 May 2003 and *Apple Daily*, 12 June 2003).

5   Current Issues in Urban Renewal: Views from Residents

The new era of urban renewal in Hong Kong is by now almost two years old. It is time to review the residents' attitudes and views towards the new urban renewal policy and their comments on the implementation of the projects. The following sections thus focus on identifying issues in urban renewal from the perspectives of the affected residents. The discussion is based mainly on data collected between February and June, 2003 by our research team, to which reference has been made at the start of this chapter. The study aims at gauging the impact of urban renewal on affected residents, and includes a survey by interviewing 539 property owners and tenants in eight projects grouped under five 'districts' for discussion purposes (see Chapter 6 in this volume for details.) We also conducted in-depth interviews with 20 residents and three focus groups in order to collect more thorough and detailed data about the views, experiences and feelings of residents during the process of redevelopment. Below we present data gleaned from our research to highlight several issues of considerable importance and, we hope, to shed light on possible improvements that can be made in future urban renewal work.

5.1   Do Residents Wish to Leave?

The stated objective of urban renewal is to improve the living conditions of residents living in old urban areas. It is assumed that affected residents would welcome the redevelopment work and look forward to moving out of their old residence. They are expected to have a strong desire to leave as the buildings are dilapidated. However our study found that quite a substantial proportion of residents do not want to leave their residence or the community: more than 50 per cent of respondents in the survey told us that they would miss their community if they were relocated.

From our interviews we found that a least 20 low-income owner occupiers living in relatively new buildings did not want to move out. More importantly, many of them expressed the view that urban renewal put them in an uncertain situation and would impose extra economic burden on them:

> To be honest, I live comfortably in this old type of building. Even in case of fire, there are sufficient safety facilities such as fire hoses and a fire alarm. It is very good for us. Secondly, I have lived in Sai Yin Pun for more than 40 years. Here I have a strong social network. Besides Sai Yin Pun is a very convenient place. I can buy what I want here. Also it is quiet. Therefore it is a very good residential environment for me.... There is quietness in this busy street.... So even if you give me a large sum of money, so what? Even if you give me a larger flat, I still don't want it. So, moving out only exerts a more adverse effect on me!

Relocation involves disruption costs and problems of adapting to a new community. Residents who focus on costs and problems, rather than on positive gains from relocation, are understandably reluctant to vacate their residence and leave the community. The more the URA knows about residents' desire to stay put or to move out, and why, the better it will be able to anticipate and minimize the scale of resistance from aggrieved residents on the one hand, and on the other hand, to take positive measures to help them to cope with any adverse impacts of relocation.

5.2   Attitudes Toward URA's Land Resumption Powers

Most of the respondents acknowledged the necessity of granting land resumption powers to the URA by the Government. However, their experience of negotiation with the URA has convinced them that such powers were one-sided and to their disadvantage. Since the URA can apply for land resumption more easily now than before, the owners' power to negotiate during the process of property acquisition has greatly diminished.

Many owner-occupiers thought that it was not fair that they did not have the right to negotiate the price with the URA. They thought that it contravened the spirit of free enterprise. Some expressed doubts that what appeared on paper as a generous compensation was in fact below the real value of a 7-year-old replacement flat, which was the agreed point of reference. These sentiments and distrust were evident from the following responses:

> The URA is a public enterprise, but it is on the side of the Government, not residents. It is very high-handed. Ordinary people have no genuine right to express their needs and negotiate the price. As a matter of fact, I have no right not to sell my property.

> I think the URA is very high-handed. It does not allow owners to negotiate the price. It dictates the purchase price. It is not up to me to sell or not sell under the URA's power to repossess my flat. Because if I don't sell it, the URA will expropriate my house. I have no right not to sell it. Even if I don't sell it now, eventually I will have to give it to URA. I cannot continue to live here… the URA has so much power.

> The Urban Renewal Authority is like a villain. One day the villain sees a girl and wants to possess her. He tells the girl's father that on 14 April, he will give him some money and take away his girl. If the father agrees, the villain will give him a living fee. If the father does not agree, the villain will take her away anyway. The father has no legal right to appeal because the villain has a royal mandate to take away the girl. Well, the girl is my flat, and I am her helpless father.

5.3   Compensation Packages

According to the guideline of compensation based on land resumption, affected owners of domestic premises are compensated with the market value of their properties plus a Home Purchase Allowance. The compensation is supposed to be sufficient to purchase a notional 7-year-old domestic flat of similar size in the same general locality. However, many respondents found that the compensation was lower than the average prices of 7-year-old to 10-year-old buildings. Some respondents could not even purchase a 20-year-old replacement flat in the same district with the compensation money. Below are illustrations of the disappointments and dilemmas felt by affected respondents:

> Suppose the URA's offered value is really a 7-year-old flat value and with this lump sum I can buy a 7-year-old flat of same size. Then it would be reasonable! However I am worried that the amount of compensation is insufficient for me to purchase a 7-year-old replacement flat of same size.... It is impossible to find such a flat in Wan Chai!

> In fact I never thought about making any profit from urban renewal. I suggest a flat-to-flat exchange. If you can give me a flat of comparable size in the same district, that would be good! And I am not bothered by all that house moving procedure. So I don't need any monetary compensation. Give me a similar sized flat and let me claim back my moving expenditure. To be honest, living in a building with no elevator does not cost me much. Suppose in the future I live in a building with elevators, the management fees will increase drastically. It will increase our economic burden.

> If the lump sum could help me buy a 7-year-old flat of the same size, I am willing to move. However, in this process I will have lost something.... Firstly, I will have to give up our community and our social network. Secondly, I will miss our neighbourhood. Honestly, by moving to another district, I will lose my school network here which, I think, is the best.

5.4   Calculation of the Notional 7-year-old Flat Value

Many owners said that they did not know how the URA evaluated the value of the 7-year-old replacement flats in the same district. The URA explained only the procedure but not the details of property appraisals to owners. It was believed that such a disclosure would stimulate more controversy over the value of a 7-year-old flat and thus would obstruct urban redevelopment. For the owners affected, the URA's arrangement failed to convince them that the compensation offered was fair.

According to the URA's property acquisition procedure, the URA first conducts a household freezing survey and registered residents as urban renewal beneficiaries. Over a period of two to three days, URA staff visits each residential unit to confirm the identity of the owner-occupier. It then recruits two surveyors to assess the notional 7-year flat value and adopts the higher value as the basis for compensation calculation. After the URA has announced the compensation offer to flat owners, owners can hire their own surveyor to re-assess the 7-years-old replacement flat value and propose a counter compensation proposal. Once both the URA and the owners have come to an agreement on compensation, the URA reimburses owners for the surveyor fees. Although a negotiation mechanism exists controversies over property appraisal and compensation still occurred.

In July 2002, it was reported that the URA had changed the property valuation method. Under the new method, the URA was required to hire seven surveyors to assess the properties. The surveyors were supposed to follow the government compensation policy on land resumption to conduct a 7-year-old property appraisal. The highest and lowest values would be disregarded. The average of the remaining five values was then taken to be the 7-year-old flat value. This is called the "weighted average method". The URA claimed that this valuation method was more objective than the old one. Yet some residents made the point that, according to a letter from the Hong Kong Institute of Surveyors to the Central and Western District Council, the Institute has never approved the "weighted average method".

5.5   Unitary Compensation Package to All Types of Buildings

Regarding a single notional 7-year-old building rate, owners living in a relatively new but small sized apartment (12 to 20 years old) criticized the URA's adoption of a unitary compensation rate for all types of buildings in the neighbourhood. For example, a female owner living with four family members in a 14-year-old flat of 350 square feet (gross floor area) in Sai Ying Pun got an offer of HK$950,000 with an extra moving allowance of HK$96,000. Despite her efforts, she could not find any flat of similar age and size in the same district. Older flats were available but these were slightly bigger (400-450 square feet of gross floor area) and would cost between HK$1.05 to 1.1 million. Without incurring extra costs, she could not find an affordable flat, let alone a 7-year-old replacement flat. Another owner-occupier living in a 15- to 20-year-old flat strongly criticized the uniform way of compensation that precluded a consideration of buildings of different ages and conditions. Besides, the majority of owner interviewees believed that the Government would not give them the amount of compensation that accurately reflected the market value of a 7-year-old flat in the same district. Instead of monetary compensation, they preferred to exchange their current flat for a 7-year-old flat in the same district in order to mitigate their financial and social losses. With regard to social losses, a respondent said, "Loss refers not just economic cost, but also to other social costs in terms of community facilities, satisfactory residential environment and social networks".

Our study found that owner-occupiers who lived in a relatively new building with good hygiene facilities and maintenance did not want to move away unless the compensation was high enough to purchase a suitable replacement flat. For those who lived in a relatively new building or who own "negative assets", the opportunity cost of surrendering the property to the Government was especially high. Thus they strongly demanded an alternative compensation policy that was different from the existing one that was based on the assumption that buildings under urban renewal were generally over 30 years old.

> Compensation aimed at purchasing a 7-year-old replacement flat certainly seems to be a good deal. Our building is 12 or 13 years old. URA's compensation criterion is uniformly applied to all types of buildings, whether they are 12 or 30-years-old. I wonder how the URA can use this unitary approach for buildings of different ages and conditions.

6 Conclusion

The contemporary era of urban renewal marks a new development in urban renewal. The URA has done a lot to help improve the living conditions of many residents in old urban areas. On paper, its work is well-intentioned, comprehensive and beneficial to the Hong Kong community; but in practice thus far there are problems. In this chapter, we have identified some of the problems from the point of view of the residents affected. Our stress on problems rather than on achievements in no way denies or undervalues the latter, but is meant to be constructive by giving a voice to residents who, for various reasons, feel aggrieved. By way of conclusion, let us return to the 'people-centred' mission of the URA.

The avowed spirit of the people-centred mission is to take good care of residents' needs and interests, to minimize the adverse impacts on them and to protect their rights in the process of redevelopment. Thus the URA did try to help owner-occupiers to find suitable replacement accommodation by ferrying them to second-hand Home Ownership Scheme flats outside renewal sites, providing them with a list of salable flats in their district and so forth. The people concerned, however, do not all see it that way. Some of their views alluding to this discrepancy have been summarized in the previous section. The more direct views are worth noting because they seem to bring out the nature of a fundamental challenge facing the URA, namely, the dilemma between its soft, people-centred mission and the harsh financial environment in which it operates. It is a tough job to be people-centred and satisfy the expectations of residents while trying to make ends meet. Understandably, URA staffs are under pressure to deliver outcomes within their financial constraints and according to schedule. Thus staffs at the front line of direct contact with residents who are distraught and aggrieved have a hard time and many of them became the focal point of severe criticism:

> Right! They brought us a long list of salable flats. I followed the information and called the sellers or agents. But some of the flats have already been sold and others were more expensive than the listed price. Actually, the URA just downloaded the information from the computer without ensuring that the information did reflect the real property market situation. Anyway, before we sold the property to the URA, the URA staff visited us regularly and were polite to me. After I sold the flat, they vanished! What they really wanted was to entice us to sell our property!

> Whenever a URA staff member called me, he just reminded me to sell the property before the deadline, or 20% of moving allowance would be deducted. Basically he did not care about our rehousing problems. He just wanted to acquire the property as soon as possible.... Now with the

> current compensation, I cannot purchase a suitable
> replacement flat, let alone a 7-year-old flat. He did not
> appreciate the difficult situation that I was in and just asked
> me to make more effort to find a new flat. Basically the
> URA does not help us to purchase 7-year-old flat in this
> district. Do you call that people-centred!

Further, according to our survey data, only 36 per cent of respondents who answered the question "Do you think that Urban Renewal Authority is on the side of residents?" agreed, whereas, in response to another question, "Do you think that Urban Renewal Authority is on the side of private developers?" as many as 72 per cent of respondents agreed.

The dilemma, we believe, affects not only the URA and the residents concerned, but also Hong Kong generally, in light of the increasing ageing of buildings now spreading all over the territory. It makes all the more sense to break out of the narrow and financially unsustainable target of redevelopment to also come to grips with the other 3 Rs (preservation, rehabilitation and revitalization). Historically, the urban renewal of Hong Kong made progress at a snail's pace. Given this, Hong Kong cannot afford to lose time in harnessing concerted efforts from stakeholders to do a good job at the 4 Rs without sacrificing the needs of residents affected. It is hoped that our research presented in this and the Ng *et al.* (2004) will help to move the 4Rs along.

References

1. Adams, D. 2000. *Urban Renewal in Hong Kong: The Record of the Land Development Corporation*. London: RICS Research Foundation.

2. *Apple Daily*, 16 June, 2002, A10. (in Chinese)

3. Fisher, S. 2001. Social Impact Assessment in Hong Kong, in Quality Evaluation Centre (ed.) *Social Impact Assessment: Asian Perspectives*. Hong Kong: Quality Evaluation Centre, City University of Hong Kong.

4. Koo, W. Y. 2001. Community Impact Analysis for Urban Renewal in Hong Kong: An Illustration of the Application of the Method. *Unpublished Master Dissertation*, Hong Kong: Centre of Urban Planning and Environmental Management, University of Hong Kong.

5. Ng, S. H., Pong, R., Kam, P. K., and Ho, C. K. 2004. The Quality of Life of Residents in Five Urban Renewal Districts. In A. Y. T. Leung (ed.) *Building Dilapidation and Rejuvenation in Hong Kong*. Hong Kong: City University Press and the Hong Kong Institute of Surveyors (BSD).

6. Planning and Lands Bureau. 2001. *People First — A Caring Approach to Urban Renewal: Urban Renewal Strategy. Hong Kong: The Printing Department*, Hong Kong Special Administrative Region Government.

7. Planning Department. 1999. *Urban Renewal Strategy Study: Executive Summary*. Hong Kong: The Printing Department, Hong Kong Special Administrative Region Government.

8. *Sing Pao*, 22 May 2002, A14. (in Chinese)

9. *Sing Tao*, 19 April 2002, A14. (in Chinese)

10. *Sun*, 26 May 2002, A13. (in Chinese)

11. Urban Renewal Authority. 2001. *Publicity Brochure*. Hong Kong: Urban Renewal Authority.

12. _____. 2002a. URA Welcomes Government's Financial Support. Retrieved 19 July 2003 from http://www.ura.org.hk/html/c602000e31b.

13. _____. 2002b. URA Achieves Significant Progress in Acquisition Projects (17 July, 2002). Retrieved July 19, 2003, Hong Kong: Urban Renewal Authority. http://www.ura.org.hk/html/c602000e39b.html

14. _____. 2003. *Work of the Urban Renewal Authority*. May, Hong Kong: Urban Renewal Authority. Retrieved from the Panel Planning, Lands and Works [CB(1)1485/02–03(04)]

Acknowledgements

The authors would like to thank Dr. Raymond Pong for his constructive comments made on an earlier version of the chapter.

4

The Economic and Social Impact of Redevelopment — A Hong Kong Case Study

K. W. Chau, Lawrence W. C. Lai, W. S. Wong, C. Y. Yiu and S. K. Wong

Only a few studies have investigated the value enhancement effects of redevelopment and rejuvenation. Redevelopment in Hong Kong has created abrupt discontinuities in originally contiguous shopping areas. Such projects have abandoned the parochial commercial and social heritage of their areas in favour of international office and professional activities. This interpretation lends support to the warning by planning theorists that urban renewal should involve broad community support and involvement. However, the rejuvenation of one large-scale estate produced positive price effects. We recommend a cost-and-benefit analysis of redevelopment versus rejuvenation in future projects. Our findings also encourage further discussion concerning the best way to promote urban renewal.

Chapter

4

The Economic and Social Impact of Redevelopment — A Hong Kong Case Study

K. W. Chau, Lawrence W. C. Lai, W. S. Wong, C. Y. Yiu and S. K. Wong

1 Introduction

Zoning is a means used by governments to control land use; thus controlling the types of urban activities carried on different pieces of land. Zoning is well-recognized by welfare economists as an effective means of tackling perceived externalities[1] arising from conflicting land-use. Some economists (such as Alexander, 1992) argue that, if zoning really is an effective tool in tackling conflict in land-use, then it should enhance property values. Whether this argument is valid or not is an empirical matter which can be verified by econometrics measures. This study empirically evaluates one zoning measure which is being used to promote urban renewal in Hong Kong.

To do this we have used the econometrics tool of hedonic pricing modelling. The result of this empirical study of the effect of zoning on land values is inconsistent with similar previous empirical studies. Past studies have generally adopted the regression method, which incorporates a given land-use category (such as commercial or residential use) as one of the independent dummy variables. However, this regression method is unable to differentiate price effects due to the proximity to a certain community facility or area from those due to the zoning measure itself. To control the effects of location, this study here makes use of the price gradients derived from a hedonic pricing model for a site zoned as Comprehensive Development Area (CDA) in Hong Kong, to identify the price effects of this zoning class.

2   Literature Review on Hedonic Pricing Modelling

The empirical results of hedonic regression on the effects of zoning on property values in most empirical studies are diverse and inconclusive. Thirty-three studies of the price effects of zoning have been reviewed in Chau et al. (2003). Twenty of these report that the effects of zoning are statistically significant (for example, Thorson, 1997; Groves and Helland, 2002), whereas 13 (for example, McMillen and McDonald 1993; McDonald, 1998) conclude that the effects are insignificant. No conclusive argument could therefore be drawn from these studies.

These empirical studies were mainly based on the experience of the North American continent and Europe. Similar studies for Hong Kong are seldom to be found. Only a few studies on the price effects of zoning in Hong Kong have been conducted (Lai, 1996, 1997). Lai (1997) attempted to evaluate the effect of zoning on price on a policy to reduce industrial plot ratios, by comparing the average price performance of private housing in a district before and after the reduction of plot ratios. No significant change after the inception of the industrial plot ratio reduction was found in the price trends of residential estates adjoining the regulated industrial area. Apart from directly measuring the impact on price, Lai (1996) also used a simple regression method to indicate that the number of planning applications and the incidence of nuisance (measured by the proxy of the number of environmental complaints) were not significantly correlated. Lai argued that the externality issue is an empirical matter, concluding that, in order to evaluate the value of effects of zoning, empirical tests should be carried out.

The class of zoning in this study is Comprehensive Development Area (CDA) zoning in urban areas. Comprehensive Development Area zoning is imposed statutorily by Outline Zoning Plans on either large sites with designated uses that are now obsolete (such as dockyards) held under single ownership or a host of small sites held under multiple ownership. In both cases, the purpose of the CDA zoning class is to make sure that urban renewal in these properties is subject to such zoning, by requiring the submission of a master layout plan for the entire zone by one development agency for the approval by the Town Planning Board. For a CDA with multiple land titles, the Urban Renewal Authority (formerly the Land Development Corporation, a government-owned corporation) is involved in such redevelopment. The Authority may spell out its powers, if necessary (under the Lands Resumption Ordinance[2], which was previously called Crown Lands Resumption Ordinance) to acquire the properties which are included in the redevelopment project.

3 Methodology

To measure the effects of CDA zoning on housing prices, a cross-sectional inter-temporal (panel) data analysis was used to identify the changes in price gradient[3] in an important CDA zone, before and after the completion of redevelopment inside the zone. A hedonic pricing model incorporating the price gradient was employed to estimate the implicit prices of the housing attributes. The refutable hypothesis for the model is that "upon comprehensive redevelopment, the CDA exerted a net positive price effect on residential properties in its neighbourhood."

In order to carry out this empirical test, a large number of transactions and their housing attributes have been obtained. The frequency of housing property transactions in Hong Kong has been high, due to the growth of the economy and high population growth. The number of transactions in the period from May 1991 to March 2001 was large enough for our analysis. These data are obtainable from a databank which is open to the public at zero cost and maintained by a major local real estate agent.

In this study, two adjoining parcels of land in Sheung Wan District that were both zoned as CDAs in the OZP in February 1990 were examined. The two areas were intended to be redeveloped into two very large-scale commercial buildings. Construction work did not, however, commence until September 1993.

The two development projects are (1) The Centre (Jubilee Street/ Queen's Road Central) and (2) The Grand Millennium Plaza (Wing Lok Street/Queen's Road Central). Figures 4.1 and 4.2 show the block plan of the area before and after the redevelopment. Figure 4.1 shows, in particular, the street shop front lost as a result of the site assembly for the redevelopment projects, which are represented in Figure 4.2.

Figure 4.1   Original Block Plan before Redevelopment (showing street level shop fronts lost as a result of redevelopment)

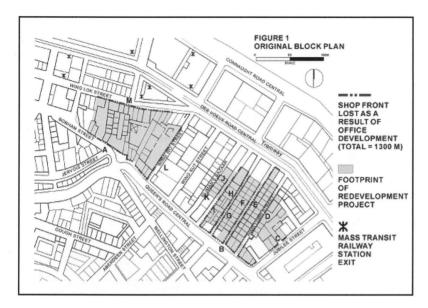

Figure 4.2   New Block Plan after Redevelopment

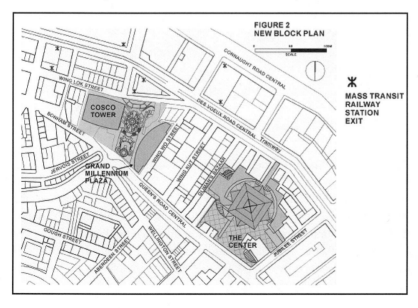

Both projects, as shown in Figure 4.2, are Land Development Corporation schemes. For both schemes, the compulsory government purchase of private properties under the Lands Resumption Ordinance and contested at court, was involved. Such properties were located at Queen's Road Central, Bonham Street, Wing Lok Street, Jubilee Street, the street blocks between Hing Lung Street and Gilman's Bazaar and Wing Wo Street. The government ignored the counter-proposal of the landowners affected but approved the master layout plans submitted to the Town Planning Board by the Corporation. The developments were completed at more or less the same time — the Grand Millennium Plaza and COSCO Tower in June 1997 and the Centre in June 1998. Figure 4.3 shows the streetscape of the Centre project, as seen today.

Figure 4.3   The Centre Abutting Queen's Road Central (See B in Figure 5.1)

The CDAs are both located in a prime location of the Sheung Wan District, are next to a mass transit railway (MTR) station, which is a major positive asset in Hong Kong where rapid public rail transport is the dominant mode of transportation. Consequently, the effects of distance from a MTR station are fully absorbed in the price gradient variable. We divided the district into six zones in order to evaluate the effects of variation in distance from the CDAs. Zone 1 is the closest to the CDAs and Zone 6 is the farthest from them.

We adopted the hedonic pricing model to evaluate whether the implementation of urban renewal policy has had any effect on price. The model is in semi-logarithm form and estimated by the Ordinary Least Squares (OLS) technique. In both models we put the age, size and storey level of a flat unit as independent variables, with the purpose of making sure that the different attributes of the different data (flat units) were controlled. The flat unit itself would be priced differently, because of the different attributes of the housing. It was expected that effect of size and storey level of a flat unit would be positively related to the housing price, but at a diminishing rate, because of the diminishing margins of returns. The higher the floor level and the larger the flat size, the more expensive the unit would be. As far as the effect of age was concerned, it was expected that the older the building, the lower the housing value, although it was uncertain as to whether this would be at a diminishing or an increasing rate.

Apart from the three independent variables that we mentioned above, the focus of the study was on the distance between the CDA zone and the flat unit. If the CDA zoning really had a net positive impact on housing price, we would expect to see that the prices of housing closer to the CDA zone (after the completion of the project) would be higher than before the project was completed.

4   Results and Interpretation

From the results of the tests, we found that age, size and floor level really do have an influence on the housing price. The effect of age and floor level are not, however, strong in the second model.

More importantly, however, in both models we found that before the development of the CDA, the farther the housing from the CDA zone, the lower its price (namely, a decreasing price trend). This is reasonable, because the CDA is sited within a prime commercial location and is relatively close to the mass transit railway stations. However, from our test result, the decreasing price trend becomes less clear. The price trend before and after the CDA was built is shown in Figure 4.4. The results cannot therefore support the hypothesis that the CDA upon comprehensive redevelopment has exerted a net positive price effect on residential properties in its neighbourhood.

Figure 4.4   Price Gradients before and after the Development of the CDAs

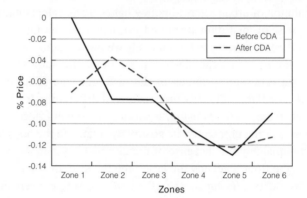

5 Conclusion

Zoning control is advocated on the basis of supposed benefits deriving from the effect of improvements in the neighbourhood, to deter the problems that arising from conflicting land use. The benefits that zoning control advocates perceive are, however, only one side of the coin. The counter-balancing effects that people often overlook are the restrictions in land use. For example, the benefits expected from zoning control on land price can be easily overwhelmed by the abrupt discontinuities in originally contiguous shopping areas. Whether there is any net benefit from zoning control is an empirical matter which requires empirical methods of evaluation. In this study, we adopted a statistical method to find out whether there was a net benefit from the completion of commercial developments in a Comprehensive Development Area zone in Hong Kong to the residential housing surrounding it.

Though our results show no enhancement in neighbouring residential property values from CDA zoning, this study focuses only on one district in Hong Kong and it should therefore not be over-generalized. In order to find out whether there is any net benefit from a CDA project, it is necessary for us to carry out a comparative study involving the examination of the value impact of major comprehensive redevelopment by developers without government assistance. It is not our intention to conclude that zoning control, such as CDA zoning, is ineffective in the planning or economic senses.

Our findings also provide insights for further discussion concerning the best way to promote urban renewal. Both the CDA projects in our study were completed without the involvement of market forces. They were the result of the involvement of a government urban renewal agency backed by the forced purchase of private properties. Our findings affirm that such government-led projects could become exclusive by failing to confer any obvious economic benefit on the adjoining properties, measured in terms of their values. A more comprehensive qualitative analysis of the master layout plans and the property management mode need to be conducted in order to explain this, but this falls outside the scope of this chapter.

The erection of the two commercial buildings of the CDA projects, unfortunately, has blocked existing neighbourhood shop frontages. A total 1,300 metres in length of shop fronts at street level has been lost as a result of the redevelopment projects. They have destroyed the existing commercial and social heritage of the area in favour of international offices and professional activities. Any urban renewal project should include broad community support and involvement. This view concurs with the idea that any proposed change in land use causes conflicts among users, as their wants are not cohesive (Kaufman and Smith, 1999). In terms of architecture, the conventional belief that the provision of "public space" (such as the vast, empty hard surfaces created at ground level in the projects evaluated), always increases the sense of community and strengthens community bonds in people in the same district may be correct only if certain design factors are satisfied (Talen, 2000). Socially speaking, the concept of "neighbourhood" (Kallus and Law-Yone, 1997) must not be neglected in urban renewal planning.

Acknowledgements

The authors would like to thank Mr. Yau Yung for collecting data for this chapter.

Notes

1. Externalities refer to the results of activities that cause incidental benefits or damages to others with no corresponding compensation provided to those who generate the externalities (Baumol and Blinder 1985, 539).

2. Cap. 124, Laws of Hong Kong.

3. Price gradient analysis is good at analysing panel data by comparing the temporal price differentials among different locations. For example, Chau

and Ng (1998) used price gradient to identify the effects of transportation improvement on housing price.

References

1. Alexander, E. R. 1992. *Approaches to Planning: Introducing Current Planning Theories, Concepts, and Issues*. Philadelphia: Gordon and Breach Science Publishers.

2. Baumol, W. J. and Blinder, A. S. 1985. *Economics: Principles and Policy*, 3rd Edition, San Diego: Harcourt Brace Jovanovich.

3. Chau, K. W. and Ng, F. F. 1998. The Effects of Improvement in Public Transportation Capacity on Residential Price Gradient in Hong Kong, *Journal of Property Valuation and Investment*, 16, 397–410.

4. Chau, K. W., Yiu, C. Y., Wong, S. K. and Lai, L. W. C. 2003. Hedonic price modelling of environmental attributes: a review of the literature and a Hong Kong case study, in LWC Lai and F Lorne (eds.) *Understanding and Implementing Sustainable Development*, New York: Nova Science.

5. Groves, J. R. and Helland, E. 2002. Zoning and the Distribution of Location Rents: An Empirical Analysis of Harris County, Texas, *Land Economics*, 78, 28–44.

6. Kallus, R. and Law-Yone, H. 1997. Neighbourhood — the Metamorphosis of an Idea, *Journal of Architectural and Planning Research*, 14, 107–125.

7. Kaufman, S. and Smith, J. 1999. Framing and Reframing in Land Use Change Conflicts, *Journal of Architectural and Planning Research*, 16, 164–180.

8. Lai, L. W. C. 1996. *Zoning and Property Rights: A Hong Kong Case Study*. Hong Kong: Hong Kong University Press.

9. Lai, L. W. C. 1997. Property Rights Justifications for Planning and a Theory of Zoning, in D. Diamond & B. H. Massam (eds), *Progress in Planning*, 48, 161–243.

10. McDonald, J. F. 1998. Land Values, Land Use, and the First Chicago Zoning Ordinance, *Journal of Real Estate Finance and Economics*, 16, 135–150.

11. McMillen, D. and McDonald, J. 1993. Could Zoning have Increased Land Values in Chicago? *Journal of Urban Economics*, 33, 167–188.

12. Talen, E. 2000. A Conceptual Model of the Relationship Between Public Space and Sense of Community, *Journal of Architectural and Planning Research*, 17, 346–360.

13. Thorson, J. A. 1997. The Effect of Zoning on Housing Construction. *Journal of Housing Economics*, 6, 81–91.

5

The Cost and Benefit of Refurbishment with Special Reference to Multi-ownership Apartment Buildings

K. W. Chau, C. Y. Yiu and S. K. Wong

It is often claimed that refurbishment (or rehabilitation) should be carried out in obsolete buildings because it can improve the buildings physically, functionally, and economically. This chapter aims to verify this claim by comparing the cost and benefit of refurbishment with special reference to multi-ownership apartment buildings. Using the property transactions data of two adjacent large-scale housing estates in Hong Kong, we found a significant net increase of property values after refurbishment. Given the growing number of aged buildings in Hong Kong, the results pose significant policy and practical implications on the building refurbishment decision process in multi-ownership apartment buildings, such as the registration of refurbishment contractors, the mandatory building inspection scheme, and the mandatory formation of incorporated owners. In particular, to drain the market force in promoting an efficient refurbishment market, the possibility of establishing a refurbishment investment market is discussed.

Chapter

The Cost and Benefit of Refurbishment with Special Reference to Multi-ownership Apartment Buildings

K. W. Chau, C. Y. Yiu and S. K. Wong

1　Introduction

It is widely recognized that buildings become obsolete and depreciate as they age. In general, the physical life of a building is so long that physical deterioration due to wear and tear is not the main reason for obsolescence, especially if periodic maintenance is in place. A more important reason is fast changing user requirements driven by technological progress and increasing competition. This leads to functional obsolescence, which refers to the inability of buildings to accommodate economically their intended new use. From a broader perspective, it is also possible that an aged building becomes socially or aesthetically incompatible with its surrounding environment because of its outdated appearance, etc. Depending on government policies and the attitude of pressure groups, the outcomes of the obsolescence can be drastically different. At one extreme, the incompatibility is perceived as undesirable, and redevelopment is urged. At the other extreme, the obsolete building is deemed historically valuable and conservation is called for.

To maintain the functionality and social status of obsolete buildings, the option that building owners usually consider is refurbishment (or rehabilitation). This chapter aims to investigate the costs and benefits of refurbishment with special reference to multi-ownership apartment buildings. If refurbishment is beneficial, then property prices should experience a net increase after refurbishment, keeping other factors unchanged. The difficulty of measurement is how to keep these other factors, such as general economic factors, unchanged. In the article, we make use of a panel-data hedonic pricing model to overcome this difficulty. The model should be useful for property managers, project consultants, and refurbishment contractors in assessing refurbishment performance. In particular, we hope that the results of the empirical study will assist the government in policy-making and owners of multi-ownership buildings in making decisions about refurbishment.

The organization of the chapter is as follows. "Previous Studies" reviews the theoretical models and empirical tests done by previous studies on refurbishment. "Methodology" proposes a panel model to measure the value of refurbishment. "Data" describes the data used for empirical analysis. "Results" reports the results and "Policy & Implications" discusses the policy and practical implications of the results, followed by concluding remarks.

Before proceeding to the next section, it is worth clarifying the meaning of refurbishment. Mansfield (2002) surveyed the meanings of refurbishment and found that different people, including professionals and policy makers, had a very different understanding about what refurbishment means. Notwithstanding these ambiguities, the Royal Institute of Chartered Surveyors' (1973) definition is good enough for our purpose, where "rehabilitation" is defined as: "… the carrying out of building work to any property, or series of properties, beyond routine maintenance, thus extending its life to provide a building or buildings which are socially desirable and economically viable."

2 Previous Studies

It has often been assumed theoretically that refurbishment enhances the market value of aged buildings because it restores and improves the physical and economic conditions of the building. The earliest theoretical works include Dildine and Massey (1974), Sweeney (1974), and Ohls (1975), who developed models to analyse the opposing effects of housing depreciation and maintenance. Arnott, *et al.* (1983) extended the models to the area of rehabilitation. Elliot, *et al.* (1985) modified Dildine and Massey's model and suggested that neighbourhood quality was one of the determinants that affected the effectiveness of maintenance. To determine the optimum level of maintenance, Vorst (1987) established a model of maintenance choices under uncertainty. The model was enhanced by incorporating the choices of upgrading, downgrading, and demolition (Anas and Arnott, 1991), and was further improved to take into account the change of value of refurbishment over time (Kutty, 1993).

Despite the abundance of theoretical work, the suggested models were rarely verified by real world data. Very few theoretical models were subject to empirical tests and most such tests were based on simulated data. For example, Wong and Norman (1994) used numerical simulations to determine the optimum timing for renovating a shopping mall. Dubin (1998) verified his discrete model of maintenance decisions by simulations. Wong (2000) adopted a numerical method to examine how sensitive the capital values of commercial properties were to refurbishment cycles.

One of the very few empirical studies that used real life data was Dunse and Jones (1998). They performed a cross-sectional study of the office market in Glasgow and found that the asking rent of a refurbished office exceeded that of a non-refurbished one. Their results also indicated that asking rent differentials between refurbished and non-refurbished offices increased with the age of the office. However, it is doubtful whether asking rents, which represent only the view of landlords, are a good proxy for genuine rental values. Moreover, the cross-sectional approach they used did not distinguish between building-specific effects (such as location) and refurbishment effects.

The above review shows that the question of whether refurbishment enhances property values remains. The answer to this question is not as straightforward as it may seem. The difficulty lies in the lack of like-with-like comparison. Intuitively, we expect a rise in property prices after refurbishment to confirm the value-added effect of refurbishment (in an inter-temporal approach). Yet this is not necessarily true because property prices are subject to the influence of many economic factors that are not fixed over time.

Another approach is the cross-sectional analysis used by Dunse and Jones (1998), which compares property prices of refurbished and non-refurbished buildings. Unfortunately, this approach suffers from the weakness that building attributes are not homogeneous and, as a result, the comparison process may be biased. This is because a refurbished building not only differs from a non-refurbished building by the quality of the improvement due to its refurbishment, but also by other intrinsic building-specific characteristics such as its location and view. In view of the deficiencies of the above methods, we introduce a panel data (cross-sectional inter-temporal) approach in the ensuing section to overcome the difficulties in measuring refurbishment value.

3   Methodology

The previous section shows that neither the inter-temporal approach nor the cross-sectional approach can give conclusive evidence on the value of refurbishment. However, a combination of these approaches, such as the panel data (cross-sectional inter-temporal) approach, can help separate refurbishment effects from other factors. This is achieved by measuring the price differentials between a refurbished building and a non-refurbished building both before and after the refurbishment. The concept is best illustrated by an example.

Suppose building A has been refurbished in time t^*. To find out whether the value of building A has increased as a result of the refurbishment, we consider all the transactions of apartment units in building A both before and after the refurbishment. Let the prices of two transactions of units transacted in time $t_0 < t^*$ and in $t_1 > t^*$ be $V(A)t_0$ and $V(A)t_1$, respectively, with the price difference being attributable to (1) the effects of refurbishment, (2) the inter-temporal movement of the property market, (3) difference in property characteristics (such as floor level or unit size) and 4) random errors.

The panel data approach is analogous to a controlled experiment. It requires another building, B, to act as a control. Building B should be chosen in such a way that it has not undergone any substantial physical and functional change (such as refurbishment) in the period concerned, but it is a close substitute of building A. In other words, buildings A and B should share similar locations, be of the same property type and serve a similar segment of the market. Given these similarities, the inter-temporal effect of the property market on building A should be more or less the same as that on building B. As a result, the transaction prices of units from the building B in time $t_0 < t^*$ and in $t_1 > t^*$ (i.e. $V(B)t_0$ and $V(B)t_1$), can be used to remove the influence of the general price change of the property market over time.

The difference in property characteristics can be taken into account by means of the hedonic pricing model. It exploits multiple regression with the property characteristics as the explanatory variables to identify the independent effects of refurbishment. The hedonic pricing model theorized by Rosen (1974) postulates that properties are a composite market of the property characteristics, and so property prices can be viewed as a summation of the implicit prices of the characteristics. The typical functional form of the model is:

$$\ln V_{it} = \alpha_0 + \sum_{j=1}^{J} \alpha_j X_{ijt} + \sum_{t=1}^{T} \beta_t D_{it} + \gamma_1 A_i \times BR_i + \gamma_2 A_i \times AR_i + \varepsilon_{it} \tag{1}$$

where V_{it} denotes the sales price of apartment unit i in period t ($i=1, \ldots, m$; $t=0, \ldots, T$); α_0 denotes the intercept; α_j denotes the estimated coefficient of each property characteristic X_{ijt} ($j=1, \ldots, J$); β_t denotes the estimated coefficient of each time dummy D_{it}; and ε_{it} is the random error. The time dummy D_{it} takes the value 1 when a transaction occurs at time t, and zero if otherwise. Note that β_0 has been set to zero for the normalization. The coefficients α_j can be interpreted as the implicit prices of property characteristics, whereas β_t is the estimated price index. The effect of refurbishment on property price can be revealed in the estimated coefficients γ_1 and γ_2, which indicate the price difference between buildings A and B before (BR) and after (AR) the refurbishment, respectively. A_i is a dummy variable which equals 1 when the property i belongs to the building A, and zero if otherwise. Similarly, BR_i and AR_i are dummy variables which equal 1 when the property i transacted before and after the refurbishment respectively.

Furthermore, the problem of random error, which may involve variations in bargaining power between sellers and buyers, can be tackled by statistical methods with sufficiently large sample of units. Fortunately, in our database we have more than 4,000 transactions for the analysis, which is reasonably sufficient to contain the problem.

4 Data

We applied the framework proposed in the previous section to analyze the refurbishment value of a popular housing estate (Estate A) in Hong Kong. Estate A has carried out comprehensive refurbishment since the end of 1998, which included the complete replacement of the tiles on the external facade and re-plumbing. Preparation works, such as the erection of scaffolding, commenced in early 1998 and the project was completed in late 1998. An adjacent housing estate, Estate *B*, was chosen to act as a control estate. Since Estates *A* and *B* are located in close proximity to each other, they share many common amenities such as bus terminals, shopping arcades, and public services. The age and height of the two estates are also similar. These justify the close substitutability between Estates *A* and *B*. Apart from this, Estate *B* has performed only routine maintenance and minor repairs during the period concerned.

Since both Estates *A* and *B* are actively-transacted estates in Hong Kong, the amount of transaction data is large. During the periods July 1991–December 1997 and July 1999–March 2001, there were 888 and 3,206 transactions in Estates *A* and *B*, respectively. The period January 1998-June 1999 was removed from the data to avoid the effect of the refurbishment announcement and the nuisance created during refurbishment work that may have affected property prices.

5 Results

The results from the hedonic pricing model analysis show that, before the refurbishment, the value of a typical flat in Estate *A* was 5.4% lower than that in Estate *B*. After the refurbishment, the value of Estate *A* overtook that of Estate *B* by 4.4%. The results are statistically significant at the 10% level. This indicates that the refurbishment has enhanced the value of Estate *A* relative to the value of Estate *B* by 9.8% (at the price level of July 1991). The validity of the hedonic pricing model is justified by its high explanatory power (R-squared value = 91%) and the correct signs obtained for all the property characteristics.

Based on the results from the hedonic pricing model analysis, a numerical analysis can be performed to evaluate whether the property value enhanced by refurbishment was justified by the refurbishment cost. Based on the price level in July 1991, the average market price of Estate *A* was HK\$2,391 per sq. ft. Since the estimated percentage increase was 9.8%, the property value enhanced by refurbishment would be HK\$ 234 per sq. ft. The direct refurbishment cost was HK\$39 per sq. ft.[1] Therefore, the net value of the enhancement was HK\$195 per sq. ft. For a typical flat of 600 sq. ft. in size, the gain from the refurbishment was HK\$117,000.

6 Policy and Practical Implications

The results showed that the property market clearly reveals the value enhancement effects of refurbishment, as the increase in value overwhelmingly exceeds the cost of the refurbishment works. However, the general reluctance of owners, especially co-owners in multi-ownership apartment buildings, to carry out refurbishment works on their buildings reflects a certain degree of "market failure".[2] There are several reasons for this gerneal attitude. First, the information cost in the procurement process and the performance evaluation process are generally expensive. Owners do not have sufficient knowledge to evaluate the tenders offered by refurbishment contractors and monitor the quality and progress of refurbishment works. Employing consultants is possible, but is usually not economically justifiable in view of the small amount of works involved. Second, the owners of apartment buildings are co-owners of the lutive lot, which they hold as tenants-in-common of indivisible shares.[3] The negotiation cost in reaching a joint decision on refurbishment among a large number of co-owners to refurbish is understandably insurmountable. Third, the value enhanced cannot be redeemed until the property is sold. However, the cost of the works has to be paid when the decision is made. When these transaction costs exceed the net benefit of refurbishment, the works will not be carried out. Although the government has introduced a Building Safety Improvement Loan Scheme since 1998 to encourage owners to carry out inspection and repair works, Yiu (2000) found that it has been poorly received. It can therefore be inferred that a lack of funding is not the main cause of the reluctance to carryout refurbishment.

This evidence of market failure implies that the reduction of transaction costs in the whole refurbishment process is conducive to the decision to refurbish. For example, the registration of contractors for minor works and the availability of refurbishment cost information would help reduce the uncertainties in cost and quality assessments. The mandatory building inspection scheme and the mandatory formation of incorporated owners in every co-owned building may also bring about a total welfare gain by reducing the negotiation and decision-making costs. Streamlining the regulatory framework to facilitate the decision-making process that binds all the owners in a co-owned building is therefore crucial.

Apart from the above regulatory means, market forces can also be explored to promote an efficient property refurbishment market. The publicity of value enhancement information, for example, can reduce uncertainties in the decision-making process among co-owners and be used as a market signal for the quality of refurbishment works. One potential use of the information is to link the contractor's fee to the estimated increase in property value after a certain date. This contract strategy provides an incentive for contractors to carryout high quality refurbishment works. Another potential use of the information is to single out refurbishment as an investment product for investors like property management companies, renovation contractors, and even fund managers. The idea is that co-owners can sell the rights to refurbish to investors, who will undertake the refurbishment work and in turn receive the property value enhanced by refurbishment. The merit of this arrangement is that the existing owners do not have to pay refurbishment costs, but instead, they get immediate cash inflow by selling the refurbishment rights. The value of the refurbishment rights can be determined by means of repeat-sales method to construct a constant quality property price index that tracks the changes in property values before and after refurbishment. (Chau, *et al*. 2003) The rights that investors (as shareholders) hold may be registered as shareholders in the Land Registry by increasing the number of shares of the building. When the trading of refurbishment rights becomes more mature, the right to refurbish can be further securitized in the form of refurbishment investment trusts, which are similar to real estate investment trusts (REIT), so as to attract a larger amount of funds to the refurbishment market.

7 Conclusion

We have presented a panel data hedonic pricing model to evaluate whether refurbishment can enhance property value. The approach has been applied to the refurbishment project of a popular housing estate in Hong Kong. Our findings show that refurbishment increased the property value of the housing estate by 9.8%. Furthermore, the property value enhancement far outweighed the direct cost of refurbishment. This means that the building owners have realized a net gain from the refurbishment of their housing estates.

It should be noted that only direct refurbishment cost (that is, the price paid to refurbishment consultants and contractors) has been considered in estimating the net gain. Other indirect costs, such as the negotiation cost among building owners in reaching the refurbishment decision, have been ignored. The difficulty in reaching a collective decision is known to be a problem in multi-ownership buildings. We hope that our proposed model can help building owners or property managers decide whether refurbishment should be carried out, and to evaluate the past performance of refurbishment contractors in the tendering process. Furthermore, we propose to explore the reduction of transaction costs and the use of market forces in promoting an efficient refurbishment market.

Notes

1.　The total direct refurbishment cost of Estate *A* was about HK$50 million at 1997 prices for a gross floor area of about 0.91 million square feet. In other words, the cost per square foot is $55. This figure is then converted to July 1991 price levels based on a discount rate of 6% p.a., which is equal to HK$39.

2.　We refer to "market failure" as a situation in which building owners would not carry out refurbishments even if they were aware of its net benefit (i.e. the increase in property values exceeds the direct refurbishment costs). This implies that there are costs (or constraints) other than direct refurbishment costs that the owners have to consider, notably transaction costs.

3.　Nield (1997).

References

1.　Anas, A. and Arnott, R. 1991. Dynamic building market equilibrium with taste heterogeneity, idiosyncratic perfect foresight, and stock conversions, *Journal of Building Economics*, 1, 2–32.

2.　Arnott, R., Davidson, R. and Pines, D. 1983. Housing Quality, Maintenance and Rehabilitation. *Review of Economic Studies*, 50, 467–494.

3.　Chau, K. W., Leung, A. Y. T., Yiu, C. Y. and Wong, S. K. 2003. Estimating the value enhancement effects of refurbishment, *Facilities*, 21(1/2), 13–19.

4.　Dildine, L. L. and Massey, F. A. 1974. Dynamic Model of Private Incentives to Housing Maintenance. *Southern Economic Journal*, 40, 631–639.

5. Dubin, R. A. 1998. Maintenance Decisions of Absentee Landlords under Uncertainty. *Journal of Housing Economics*, 7, 144–164.

6. Dunse, N. and Jones, C. 1998. A hedonic price model of office rents, *Journal of Property Valuation and Investment*, 16(3), 297–312.

7. Elliott, D. S., Quinn, M. A. and Mendelson, R. E. 1985. Maintenance behaviour of large-scale landlords and theories of neighbourhood succession, *AREUEA Journal*, 13, 424–445.

8. Kutty, N. K. 1995. A Dynamic Model of Landlord Reinvestment Behavior. *Journal of Urban Economics*, 37, 212–237.

9. Mansfield, J. R. 2002. What's in a Name? Complexities in the Definition of Refurbishment, *Property Management*, 20(1), 23–30.

10. Nield, S. 1997. Hong Kong Land Law, 2nd Edition, Hong Kong: Longman.

11. Ohls, J. C. 1975. Public Policy towards Low Income Housing and Filtering in Housing Markets. *Journal of Urban Economics*, 2, 144–171.

12. Rosen, S. 1974. Hedonic Prices and Implicit Markets: Product Differentiation in Pure Competition. *Journal of Political Economy*, 82, 34–55.

13. Royal Institution of Chartered Surveyors. 1973. *The Rehabilitation of Homes and Other Buildings*. London: Royal Institution of Chartered Surveyors.

14. Sweeney, J. L. 1974. Quality, Commodity Hierarchies, and Housing Market. *Econometrica*, 42(1), 147–167.

15. Vorst, A. C. F. 1987. Optimal Housing Maintenance under Uncertainty. *Journal of Urban Economics*, 21, 209–227.

16. Wong, K. C. 2000. Valuing the Refurbishment Cycle. *Property Management*, 18(1), 16–24.

17. Wong, K. C. and Norman, G. 1994. The Optimal Time of Renovating a Mall. *Journal of Real Estate Research*, 9(1), 33–47.

18. Yiu, C. Y. 2000. The Effects of Ageing Buildings on Maintenance Costs in Hong Kong, *Unpublished MPhil Thesis*, Department of Real Estate and Construction, The University of Hong Kong.

6

The Quality of Life of Residents in Five Urban Renewal Districts

Sik Hung Ng, Raymond W. Pong, P. K. Kam and Charles C. K. Ho

In order to understand residents' subjective, human responses to urban renewal, two sets of housing-based Quality of Life (QoL) indices have been developed to assess the quality of the immediate living environment and of the wider neighbourhood/ community. These indices are moderately intercorrelated within each set, and weakly intercorrelated between the two sets of indices. This pattern of correlations indicates a satisfactory degree of the convergent and divergent validity of the indices, in addition to the face validity of the indices. Thus the indices provide the necessary tools for applying a QoL approach to urban renewal.

Chapter

The Quality of Life of Residents in Five Urban Renewal Districts

Sik Hung Ng, Raymond W. Pong, P. K. Kam and Charles C. K. Ho

1 Housing and Quality of Life

When the Government designates a particular cluster of buildings (henceforth referred to as a "district") for urban renewal, it makes the assumption that the buildings' physical condition has deteriorated beyond reasonable repair. Objective data on the structural condition of these buildings will no doubt inform the decision-making process. What have been overlooked, even ignored, are the residents' subjective perceptions of and (dis)satisfaction with the physical condition of their residence, and how those perceptions and (dis)satisfaction may vary from one urban renewal district to another. Knowledge of the subjective, human side to urban renewal can throw light on the situation beyond the structural conditions of buildings, particularly in the difficult and often protracted process of persuading residents to vacate their homes. For example, the Urban Renewal Authority's recent attempts at land resumption and property acquisition for development have shown that residents who are satisfied with their residential situation, compared to those who are not, are more resistant to vacating their homes (Ho *et al.*, 2003).

The human side of urban renewal is not confined to the residents' subjective perception of their immediate living environment. It also extends to the surrounding neighbourhood and community environment:

how residents perceive their neighbours and the surrounding community, the extent to which they are satisfied or dissatisfied with them, their access to interpersonal networks and social services therein, their social participation in the area and so forth. Residents can see in their "old" buildings and "old" communities economic advantages, convenience and charm that can easily escape the attention of planners and, along with this oversight, the many problems that residents have to deal with once they embark on their journey of relocation. Knowledge of residents' views and feelings would better inform our understanding of what it really means to be relocated; the social costs to individual residents, the disruption of the "social capital" (Putnam, 2000) that has been accumulated in the community, the problem of social integration in the new community and so on.

Research in the social sciences has shown that relocation or resettlement will disrupt not only the living pattern to which residents have become habituated, but also their existing social support network. Both the Organization of Economic Co-operation and Development (OECD) and the World Bank have incorporated these psycho-social considerations in their guidelines on "voluntary resettlement", a subject not unlike relocation in the present context of urban renewal and, more generally, on the importance of nuturing and retaining social capital (OECD, 2001; World Bank, 2001). In Hong Kong, the Urban Renewal Authority, set up by the government to be responsible for the massive programme of urban renewal, has similarly endorsed the importance of human considerations. This is enshrined in its "people-first" or "people-centred" policy discussed in Chapter 3 in the present volume. (See also Fisher, 2001).

In this chapter, we develop a Quality of Life (QoL) approach to understanding the housing situation of residents who are caught up in urban renewal. QoL is a multi-faceted, complex concept that has generated numerous studies ranging from sociology (Schuessler & Fisher, 1985) to medical sociology (Farquhar, 1994) and nursing (Draper, 1997), and from ageing (Abeles *et al.*, 1994) to the frail elderly (Birren *et al.*, 1991) and health care (Nordenfelt, 1994). It has become a scholarly concern on the international scene (Orley & Kuyken, 1994). To our knowledge, no QoL study has been carried out specifically on residents facing urban renewal. In this context, our research represents a pioneering attempt to extend QoL research to urban renewal, an important socio-economic development in Hong Kong and, in so doing, to shed new light on current discussion of urban renewal from a QoL perspective. Given the importance of housing as both an economic and a social concern, the usefulness of housing-based

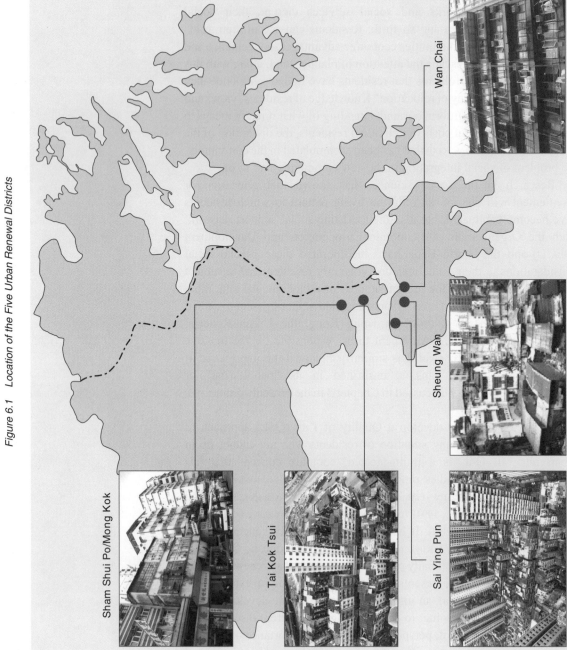

Figure 6.1   Location of the Five Urban Renewal Districts

QoL indicators to gauging social development in Hong Kong should be obvious (Hong Kong Council of Social Service, 2002).

In the report below which is based on data from a larger on-going study, we treat residents' perceptions of and satisfaction with their (a) home and (b) the wider neighbourhood/community as two housing-specific domains of QoL. Indicators in these two domains will be constructed to gauge residents' QoL and to compare the QoL of residents in five urban renewal districts. In addition to housing-specific QoL, we also adopt a generalized measure of QoL to establish the extent of correlation between housing-specific and generalized measures of QoL. For present purposes, a shorthand version of the latter in the form of "Life Satisfaction" (LS) will suffice. That is, we shall make use of LS as a simplified measure of QoL: the higher the LS the better the QoL (Neugarten *et al.*, 1961).

2 Survey

A survey involving 539 residents in five districts was carried out From February to March 2003, and later in June 2003. (The timing of the survey was determined in part by the urban renewal timetable of the Urban Renewal Authority and by the outbreak of the "Atypical Pneumonia" epidemic from mid-March to early June.) All household units still living in the districts at the time of the survey were contacted, with the aim of interviewing a household member in one of these classes: elderly (56 years or above), middle-aged (30–55 years), or young (12–29 years). Missed calls were revisited once before discarding the subjects from the sample. The response rate, excluding missed calls, was high (82%).

The 539 residents were distributed in the five districts as follows: Tai Kok Tsui (128), Sham Shui Po/Mongkok (112), Wan Chai (42), Sai Ying Pun (169) and Sheung Wan (88). Mongkok, with relatively few residents (*n*=34), was combined with Sham Shui Po as one district for data analysis because of their geographical proximity and the similarity of the structural conditions of their buildings. The relatively small sample from Wan Chai was due to the small number of affected buildings and the early relocation of the residents before the present survey began.

As can be seen in Table 6.1, for the sample as a whole, men and middle-aged residents were over-represented whereas women and younger residents were under-represented. Gender and age stratification was uneven in the five districts. To control for possibly confounding effects due to the uneven distributions of gender and age, gender and age were treated as covariates in the Analyses of Covariance (ANCOVA) below.

Table 6.1 Age and Gender Distribution of the Sample

	Gender		
	Male	Female	Total
Below 30 Year	47	41	88
30–55	142	111	253
56 and above	127	71	198
Total	316	223	539

3 Perception of and Satisfaction with the Residence

3.1 Perceived Building Problems

A perceived building problem (PBP) index was constructed by asking residents to check problems against seven categories:

1. Outside wall (such as peeling of paints and similar minor problems)
2. Inside wall and structure (such as cracks)
3. Water seeping into the building
4. Blockage of water and drainage pipes
5. Inadequate fire safety provisions
6. Unhealthy living environment
7. Other problems

Seventeen per cent of residents (n=91) indicated they had no problem at all. Another 50% encountered problems in three or more (3+) categories. If we take the 3+ PBP Index as an indication of severe problems, then approximately half of the overall sample perceived their living quarters to be in a seriously bad physical condition.

Variations of PBP Index across the five districts were tested by Analysis of Covariance (ANCOVA) with gender and age treated as covariates. The results showed a significant ($F(4, 532)=23.567$, $p<0.001$)

variation with the worst PBP Index in Tai Kok Tsui and the best PBP Index in Sai Ying Pun (see Table 6.2).

Table 6.2   Mean Perceived Building Problem (PBP) Index

Districts	PBP Index
Tai Kok Tsui	3.8
Sheung Wan	2.8
Sham Shui Po/Mong Kok	2.6
Wan Chai	2.4
Sai Ying Pun	1.8

Note: PBP Index varies from 0 (best) to 7 (worst).

3.2   Satisfaction with Residence

Residents were asked to indicate their level of satisfaction and dissatisfaction with various aspects of their residence, such as floor area, ventilation, hygienic conditions and natural light. Afterwards, they indicated their overall level of satisfaction on a 5-point scale (5 = very satisfied). For the sample as a whole, the overall satisfaction was 2.9, which was slightly on the dissatisfaction side. Analysis of covariance across districts showed a significant ($F(4, 532) = 4.873$, $p<0.001$) variation with the highest satisfaction in Sai Ying Pun and Sheung Wan, and the lowest satisfaction in Sham Shui Po/Mongkok (see Table 6.3).

Table 6.3   Mean Residential Satisfaction

Districts	Satisfaction with Residence
Sai Ying Pun	3.2
Sheung Wan	3.2
Tai Kok Tsui	2.9
Wan Chai	2.8
Sham Shui Po/Mong Kok	2.8

Note: Mean satisfaction varies from 1 (lowest) to 5 (highest).

3.3   Residential Needs

A Residential Needs (RN) Index was constructed by asking respondents to indicate what aspects of their residential conditions they would most strongly expect to improve in their new residence. They could check up to nine items, as follows:

- Size of total residential area
- Size of sleeping area
- Size of kitchen area
- Size of dining area
- Size of study area
- Size of toilet area
- Ventilation
- Hygiene
- Natural lighting

The mean level of needs varied significantly across districts ($F(4, 532) = 2.385$, $p<0.05$), with the highest need in Sheung Wan and the lowest need in Tai Kok Tsui (Table 6.4).

Table 6.4   Mean Residential Needs

Districts	Residential Needs
Sheung Wan	5.6
Sai Ying Pun	4.8
Wan Chai	4.8
Sham Shui Po/Mong Kok	4.7
Tai Kok Tsui	4.2

Note: Needs vary from 0 to 9.

Residents' confidence in realizing their housing needs was quite high with an overall mean of 3.4 (1 = no confidence, 5 = very confident). The ANCOVA results showed that the confidence levels did not vary significantly across districts.

4   The Broader Context: Neighbourhood and Community

Satisfaction (or dissatisfaction) with the immediate living environment, as reported above, is an important aspect of residents' quality of life and should not be treated as synonymous with the adequacy or soundness of the physical structure of a building, even though the latter may contribute greatly to subjective satisfaction. The neighbourhood and the community may also play an important role in shaping personal views on the living environment since individual homes are embedded in a neighbourhood, which, in turn, is part of a larger community. It is also possible that people may assign different levels of importance to different aspects of the living environment. In terms of the level of satisfaction, some may focus exclusively on the units in which they reside, while others may look beyond their homes. These are issues that need to be determined empirically.

Although urban renewal is often based on the belief that the health and safety of residents may be compromised as a result of the progressive deterioration of residential buildings and the related infrastructure, the willingness or reluctance of the residents to relocate may be determined not just by the perceived "livability" of their homes, but also by their attachment to their neighbours and the community. According to data collected in the survey, the respondents on average had lived in the same residential unit for nearly 15 years. On average, they had lived in the same district for 24 years. Social networks, human relationships and psychological attachments could have developed or emerged over those years (Moorer & Suurmeijer, 2001). Thus, it is important to assess the contextual effects of neighbourhood and community on people's willingness to co-operate with urban renewal efforts and to relocate.

4.1   Satisfaction with Neighbourhood

In addition to asking various questions about how residents felt about the physical aspects of their residences, the survey assessed the extent to which the respondents were satisfied with their neighbours and the community. It appears that the residents were generally satisfied with their neighbours. Almost 65% of the respondents said that they were "very satisfied" or "fairly satisfied" with their neighbours. Less than 8% were either "dissatisfied" or "very dissatisfied". Table 6.5 shows the mean levels of neighbourhood satisfaction in the five districts.

Table 6.5   Mean Neighbourhood Satisfaction

Districts	Neighbourhood Satisfaction
Wan Chai	3.9
Sheung Wan	3.8
Sham Shui Po/Mong Kok	3.8
Sai Ying Pun	3.7
Tai Kok Tsui	3.7

Note: Mean satisfaction varies from 1 (low) to 5 (high).

This fairly high level of satisfaction was generally borne out by three other questions on interactions with neighbours. For instance, about 36% of the respondents indicated that they talked to their neighbours "always" or "often". With respect to neighbours helping one another, 28.5% of the respondents indicated that they and their neighbours "always" and "often" helped one another. Thus, it is not surprising that 65% the respondents reported that they were "very satisfied" or "fairly satisfied" with their overall relationship with their neighbours. All three questions were fairly highly correlated with neighbourhood satisfaction.[1]

Variations of neighbourhood satisfaction across districts were tested by an analysis of covariance. The results showed no significant variation across districts.

4.2   Satisfaction with the Community

Generally speaking, the level of satisfaction with the communities in which the residents of the five urban renewal projects lived was very high. Slightly over 70% of the respondents indicated that they were either "very satisfied" or "fairly satisfied" with their community. Only about 9% of the respondents were "dissatisfied" or "very dissatisfied" with their community. Variation in community satisfaction level across districts was significant ($F(4, 528) = 8.79$, $p<0.001$), with the lowest level of satisfaction in Tai Kok Tsui and the highest in Wan Chai (see Table 6.6).

Table 6.6   Mean Community Satisfaction

Districts	Satisfaction with Community
Wan Chai	4.3
Sai Ying Pun	3.9
Sheung Wan	3.9
Sham Shui Po/Mong Kok	3.8
Tai Kok Tsui	3.5

Note: Mean satisfaction varies from 1 (low) to 5 (high).

4.3   Attachment to the Community

An indication of community attachment is whether or not the respondents would miss the community in which they live once they were "resettled". The results show that the respondents were fairly evenly split. About 48% of the respondents said that they would not miss their present community at all or would miss it "a little bit". On the other hand, about 52% said that they would "strongly miss" or would miss their present community "quite a bit". Analysis of covariance showed no significant variation across the five districts.

The proportion of respondents who would miss their community was somewhat smaller than the proportion of respondents who were satisfied with their community. This apparent discrepancy could be due to the fact that many residents believed the new community where they were to be resettled would be better than the one they were residing in at the time of the survey. Thus, about 54% of the respondents were either "very confident" or "fairly confident" that their future community would be better than their current one. It is worth noting that there was a negative correlation ($r = -0.23$; $p<0.01$) between community satisfaction and confidence that future community would be better. In other words, those who said that they would not miss their present community were more likely to believe that they would be moving into a better community.

5 Linking Residence with the Wider Neighbourhood and Community

The three residence-based QoL indices (PBP index, satisfaction with residence, and residential needs) were significantly inter-correlated, as were the three neighbourhood/community-based QoL indices (satisfaction with neighbours, satisfaction with community, and miss community). As shown in Table 6.7, the two sets of inter-correlations, highlighted respectively in italics (residence-based indices) or bold type (neighbourhood/community-based indices), varied from .17 to .38 (in absolute values) and thus indicated a moderate level of internal consistency. Between the two sets of indices, as can be expected, the correlations (shown in plain type) were generally weaker and some were not significant.

Table 6.7   *Intercorrelations among Housing-based QoL Indices*

	General Building Problems	Satisfaction with Residence Environment	Residential Need	Satisfaction with Neighbours	Do you miss your community?
Satisfaction with Community	-0.24**	0.34**	-0.14**	**0.27****	**0.30****
Satisfaction with Neighbours					**0.17****
Perceived Building Problems Index		*-0.38***	*0.18***	-0.14**	-0.06
Satisfaction with Residence			*-0.37***	0.22**	0.17**
Residential Need				-0.11*	-0.01

Note: Italicized correlations are correlations among residence-based indices; those in bold type are correlations among neighbourhood/community-based indices. Correlations between residence- and neighbourhood/community-based indices are shown in plain type.

6   Life Satisfaction

In the survey, residents were also asked to respond to items measuring life satisfaction in general:

- I am as happy as when I was younger.
- My life could be happier than it is now (reversed coded).
- These are the best years of my life.
- The things I do are as interesting to me as they ever were.

The four items were selected from the Life Satisfaction Index Form A (Neugarten *et al.*, 1961). The index is easy to administer and acceptable to a wide age range, including older people (James & Davis, 1986), thus making it a more suitable tool for the present study than other, more lengthy instruments (such as World Health Organization, 1998). The four items were translated into Chinese for the present survey and found to have satisfactory reliability (alpha = .72). To test whether QoL varied across districts, a one-way ANCOVA was carried out on the LS scores with gender and age treated as covariates. The results showed no signification variation.

6.1 Predicting life Satisfaction from Housing-based QoL Indices

The 4-item Life Satisfaction scale scores were regressed on the six housing-based QoL variables and, in order to control for demographic variations, also on gender, age, and district. The variables were coded as follows:

- Gender (male = 1, female = 2).
- Age (young =1, middle-aged = 2, elderly = 3).
- Districts (Tai Kok Tsui = 1, Sham Shui Po/Mong Kok = 2, Wan Chai = 3, Sai Ying Pun = 4, Sheung Wan = 5).
- Perceptions of/satisfaction with residence (perceived building problem index, overall satisfaction with residence, and residential needs index).
- Perceptions of/satisfaction with neighbourhood/community (satisfaction with neighbours, overall satisfaction with community, extent of missing community if relocated).

The results are shown in Table 6.8. Collectively the nine predictor variables accounted for 11% of the variance. However, only three variables had significant independent effects on life satisfaction. LS was higher (a) for men than for women, (b) when residents perceived fewer building problems and (c) when residents were more satisfied with their living quarters. Note that none of the neighbourhood/community variables had any significant independent effects, whereas two of the residence variables had.

Table 6.8   Outcome of Regression Analysis Predicting Life Satisfaction

Predictor Variables	ß	$p =$
Age Group	0.02	0.663
Gender	-0.10*	0.040
Renewal Districts	0.01	0.832
Perceived Building Problems	-0.12*	0.022
Satisfaction with Residence	0.19*	0.001
Residence Need	0.07	0.184
Satisfaction with Neifhbours (5 = very satisfied to 1 = very dissatisfied)	0.08	0.123
Satisfaction with Community (5 = very satisfied to 1 = very dissatisfied)	0.07	0.223
Extent of mssing community when relocated (4 = strongly missed to 1 = missed)	0.01	0.901

Note: * p< .05

7 Conclusion

In an attempt to reveal residents' subjective, human responses to urban renewal, two sets of housing-based Quality of Life indices have been developed to tap into the quality of the immediate living environment and of the wider neighbourhood/community, respectively. These indices are, as expected, more strongly inter-correlated within each set than between sets. This pattern of correlations indicates a satisfactory degree of the convergent and divergent validity of the indices, in addition to the face validity of the indices. Thus, the indices provide the necessary tools for applying a QoL approach to urban renewal. From a survey of 539 residents based on these indices, two broad findings are discernable.

First, QoL pertaining to the immediate living environment (residence) was generally poorer than QoL pertaining to the wider environment. For example, satisfaction with the residence (mean = 2.9) was much lower than satisfaction with the community (mean = 3.8). An important implication of this comparative finding is that urban renewal in the form of pulling down buildings and dispersing residents from the community may place residents in a dilemma. On the one hand, relocating residents to a physically better residence would improve their currently low residence-based QoL, and hence would be able to make an immediate positive impact on a domain of QoL that is most in need of improvement. On the other hand, however, moving away from a familiar neighbourhood and community that has contributed positively to residents' QoL would place them in a new, uncertain environment that may turn out to be of poorer quality. How residents cope with relocation and improve their QoL in the new environment are issues that the Urban Renewal Authority should address under its "people-centred" approach to urban renewal. It is here that social science research can play an important role in getting a better understanding of residents' coping strategies and the environmental determinants of their QoL.

The second broad finding is that in most housing-based QoL indices, Tai Kok Tsui stood out as the poorest among the five districts. If we take Tai Kok Tsui as a benchmark for identifying districts for priority urban renewal, then the housing-based QoL indices developed in the present survey may be used to prioritize the districts that are in the queue. This application is pertinent given the economic downturn in Hong Kong and the increasing need to reduce the financial burden of the Urban Renewal

Authority by taking some of the districts off the renewal list for alternative treatment such as rehabilitation and better maintenance (see Kam *et al.*, 2004). The question of how to identify districts that would be most suitable for, and would benefit the most from, building rehabilitation and other means of lengthening their life span has become critically important. The application of QoL indices, we propose, would be a viable and relatively cheap way of collecting relevant "people-centred" data for making informed decisions. Another advantage of applying a QoL approach to this issue is that decisions based on QoL would be more acceptable to residents concerned. Further, QoL information would be useful for channelling limited resources to specific areas of the residential and community environments most in need.

Having developed a QoL approach to ageing buildings by constructing housing-based QoL indices, we also attempt to link these indices to traditional QoL research by relating the housing-based QoL indices to a generalized measure of QoL (life satisfaction). The latter was regressed on the six housing-based QoL indices along with gender, age, and district. The results showed that QoL pertaining to residence, but not QoL pertaining to the wider environment, significantly predicted life satisfaction. Specifically, perceived building problems and the overall satisfaction with residence were the significant housing-based QoL predictors (along with gender). The finding suggests that at least for the present sample of residents, their immediate living environment was more important than the wider environment for their life satisfaction.

Moreover, the present survey provides a baseline prior to relocation for a future longitudinal study on residents' post-relocation QoL. The comparative data, when available, would be useful for assessing the social impact of urban renewal and, just as important, for understanding how residents cope with and make the best use of the experience, and why some of them may have been more successful than others. These are important because although the impact of rehousing may appear on the surface to be a potentially positive change to poor physical living conditions, it is in fact complex. For example, in a pioneering study of the impacts of housing the poor in Hong Kong in the 1960s, Hopkins (1971) has cast doubt on the belief that resettled people would necessarily experience an improved quality of life that was beyond the reach of squatters who remained unresettled. The current situation, some 40 years later, is even more complex. As a result, the extent to which urban renewal can achieve the Urban Renewal Authority's "people-centred" objective should not be taken for granted, but requires even closer research scrutiny.

Since people work and live in buildings, the environment created by buildings is a major determinant of their quality of life. The building and quality of life nexus, being so strongly bonded, beckons the attention of both the building and social sciences. To this end, the purpose of this chapter has been to look at the quality of life end of the nexus. Finally, let us touch base with the wider social context in which urban renewal has become a necessary and regular part of city life. As the Hong Kong population grows, so does the number of new buildings and, more importantly, the number of ageing buildings as this is the focus of the present volume. The stock of ageing buildings is vast and increasing. Many of them are falling into dilapidated and irredeemable conditions; yet, a large number of them can be rejuvenated. In the past, urban renewal policy driven mainly by economic considerations has narrowed the option of urban renewal to one of demolishing old buildings and replacing them with new ones, when in fact a wider range of options such as rehabilitation and preservation are available. These alternatives may contribute more to the qualify of life, and if so, their adoption will pay dividends. In the present climate of economic downturn and falling property prices, where demolition and rebuilding are no longer as economically attractive as before, it is all the more compelling to place the building and quality of life nexus in the wider context of rehabilitation and preservation.

Notes

1. Correlation coefficients of "satisfaction with neighbourhood" with "talked to neighbours", "neighbours helping one another" and "satisfied with overall relationship with neighbours" are 0.41, 0.38 and 0.48, respectively. All are statistically significant at $p<0.01$.

References

1. Abeles, R. P. Gift, H. C. and Ory, M. G. 1994. *Aging and Quality of Life*. New York: Springer.

2. Birren, J. E., Lubben, J. E., Rowe, J. C. and Deutchman, D. E. 1991. *The Concept and Measurement of Quality of Life in the Frail Elderly*, San Diego, CA: Academic Press.

3. Draper, P. 1997. *Nursing Perspectives on the Quality of Life*. London: Routledge.

4. Farquhar, M. 1994. Quality of Life in Older People. *Advances in Medical Sociology*, 5, 139–158.

5. Fisher, S. 2001. Social Impact Assessment in Hong Kong in Quality Evaluation Centre (ed.). *Social Impact Assessment: Asian Perspectives*. Quality Evaluation Centre, City University of Hong Kong, 26–32.

6. Ho, C. K., Ng, S. H. and Kam, P. K. 何俊傑，伍錫洪，甘炳光. 2003. Urban Renewal: Implementation Problems of Property Acquisition. 市區重建：物業收購的執行問題. *Hong Kong Economic Journal*, 137, 33–37.《信報月刊》.

7. Hong Kong Council of Social Service. 香港社會服務聯會. 2002. Five Year Social Development Plan Submitted to the Hong Kong Special Administrative Region Government. Hong Kong: Hong Kong Council of Social Service.《香港社會服務聯會向特區政府建議的 5 年社會發展計劃》.

8. Hopkins, K. 1971. Housing the Poor, in K. Hopkins (ed.), *Hong Kong: The Industrial Colony*. Hong Kong: Oxford University Press, 271–335.

9. James, O. and Davies, A. D. M. 1986. The Life Satisfaction Index — Well-being: Its Internal Reliability and Factorial Composition. *British Journal of Psychiatry*, 149, 647–650.

10. Kam, P. K., Ng, S. H. and Ho, C. K. 2004. Urban Renewal in Hong Kong: Historical Development and Current Issues, in A. Y. T. Leung (ed.) *Building Dilapidation and Rejuvenation in Hong Kong*. Joint Imprint by the CityU Press and the Hong Kong Institute of Surveyors (BSD), Hong Kong.

11. Moorer, P. and Suurmeijer, T. P. B. M. 2001. The Effects of Neighbourhoods on Size of Social Network and the Elderly and Loneliness: A Multi-Level Approach. *Urban Studies*, 38, 105–118.

12. Neugarten, B., Havighurst, R. and Tobin, S. 1961. The Measurement of Life Satisfaction. *Journal of Gerontology*, 16, 134–143.

13. Nordenfelt, L. 1994. *Concepts and Measurement of Quality of Life in Health Care*. Dordrecht, Netherlands: Kluwer Academic.

14. OECD. 2001. *The Well-Being of Nations: The Role of Human and Social Capital*. Paris: Organization of Economic Co-operation and Development.

15. Orley, J. and Kuyken, W. (eds.) 1994. *Quality of Life Assessment: International Perspectives*. Heidelberg: Springer Verlag.

16. Putnam, R. 2000. *Bowling Alone: The Collapse and Revival of American Community*. New York: Simon & Schuster.

17. Schuessler, K. F. and Fisher, G. A. 1985. Quality of Life Research and Sociology. *Annual Review of Sociology*, 11, 129–149.

18. World Bank. 2001. Social Capital Home Page, June 2001. www.worldbank.org/poverty/scapital/index.htm

19. World Health Organization. 1998. Quality of Life Assessment: Development and General Psychometric Properties. *Social Science and Medicine*, 46, 1569–1585.

7

Criteria and Weighting of a Value Age Index for Residential Use

Daniel C. W. Ho, C. M. Tam and C. Y. Yiu

Most theoretical and empirical studies regard building age as the most important criterion in property value determination. In this chapter, we have constructed a more general framework to identify the multiple criteria of value determination. The objective is to evaluate the effects of different residential building quality attributes on building values, including presentation, management, structural and system defects, building services, access and circulation, and the provision of amenities. Our findings indicate that building age is just one of the less important criteria that people consider in the evaluation of building value, while the structural and system defects of a building are found to be more important.

Chapter

7

Criteria and Weighting of a Value Age Index for Residential Use

Daniel C. W. Ho, C. M. Tam and C. Y. Yiu

1 Introduction

Building quality assessment (BQA) is one of the most important tools in modern facility management. It is used as a vehicle for assessing the adequacy of existing facilities to meet the changing needs of an organization (Lawrence and Pande, 2000). BQA is also an evaluation of the physical condition and functional performance of facilities (Kaiser, 1989).

While almost all BQA schemes are designed for commercial and office buildings only (Rider Hunt Group, 1991; Davis, et al., 1993a, 1993b; Issacs, et al., 1994; Baird, et al., 1996; Ho, 2000; So and Wong, 2002, etc.), the residential sector is the dominant portion (70%) of property assets in Hong Kong, as shown in Figure 3.1 in Chapter 3. Since BQA schemes for office buildings cannot be directly applied to residential buildings, it is necessary to formulate a specific BQA system for residential use.

2 Literature Review

Clift (1996) defined the quality of a building as "the degree to which the design and specification meets the requirements for that building". However, the requirements of users differ substantially from one building type to another. In order to build up a useful BQA index, we must identify the specific requirements of the building type to be studied. The requirements of offices and residential apartments are obviously quite different. The criteria chosen in the evaluation tools should therefore be able to cater to different uses, rather than attempt to provide a "universal" rating tool. In other words, index construction should be "building-type specific" (Soebarto and Williamson, 2001).

A BQA scheme usually covers a wide range of elements that affect the performance of a building. The BQA index was developed as a tool for scoring the quality of the multiple attributes of a building by relating its actual performance in fulfilling the requirements of the user groups. For example, Davis, *et al.* (1993a, 1993b) put forward their serviceability tools and method (STM), which assesses over 100 sub-criteria under three headings: (1) group and individual effectiveness, (2) property and management, and (3) laws, codes, and regulations. On the other hand, the Rider Hunt Group (1991) commissioned the development and made use of another BQA scheme, which was explained in detail in Isaacs, *et al.* (1994). They classified performance measurement into nine categories: presentation, space functionality, access and circulation, amenities, business services, working environment, health and safety, structural considerations, and building operations. But the massive scale of measurement makes them highly difficult to manage. Ho (1999, 2000) simplified the above models and derived a 6 × 5 criterion matrix. The six criteria are (1) presentation, (2) management, (3) functionality, (4) services, (5) access and circulation, and (6) amenities. These categories of quality indicators form the basis of the criterion framework in this paper.

These models are typically tailored towards office or commercial properties. Most of the schemes put a great deal of emphasis on energy efficiency and intelligent building services provision. This is understandable, as the running costs of office and commercial buildings are largely for building services and their users are profit-oriented. In residential flats, however, natural lighting and ventilation are more commonly exploited and the objectives of the users of these properties differ widely from those of office and commercial users. It is therefore not appropriate to apply these schemes directly to residential properties. So and Wong (2002) tried to put different weights on their nine modules' model in an attempt to provide a universal BQA scheme. Nevertheless, not only is the weight allocation arbitrary, but many sub-criteria are also irrelevant in a residential property setting. This chapter intends to develop a BQA index model specifically for residential property in Hong Kong.

3 Model

Residential property in Hong Kong is predominantly high-rise with relatively small-sized flats. Each of these buildings is commonly co-owned by a large number of owners. The maintenance of common areas and common facilities is often a difficult task (Lai & Ho, 2001). It is understandable that potential buyers would try to avoid this complication. Since it is difficult for the general public to observe these maintenance needs, the general practice in the market is to take building age as a proxy. Therefore, the value drop for aged buildings is huge. In order to understand more accurately how a building's age affects its value, 'building age' is incorporated into the model together with the other maintenance related criteria. Furthermore, the result of the BQA is coined as a Value Age Index (VAI) of residential property.

3.1 Assumptions

First, we will restrict our attention to residential property alone as their ageing and dilapidation problems demand urgent attention. Fortunately, the variety of forms and construction of residential buildings in Hong Kong is very limited. Most of the buildings are multi-storey reinforced concrete framed structures, and are co-owned by multiple owners. We will also assume that no unauthorized building works (UBWs) exist (Lai & Ho, 2001); otherwise structural safety and integrity become impossible to assess. We will also try to minimize the number of criteria to make the model more feasible, and yet accurate enough to differentiate various levels of quality achieved. We have therefore included only major attributes of property-specific quality in the model. Consequently, spatial functionality and location factors, among others, are excluded. As the minimum requirements of regulations stipulating the health and safety design of residential buildings in Hong Kong are strictly followed in almost all buildings, the criterion of basic health and safety is also eliminated in this study. The model developed assesses only those areas that are commonly owned, as it is impractical to gain access to the areas within the exclusive possession of individual owners. Finally, the initial quality of construction is ignored as it is believed to be highly correlated with the structural and system defects, as well as the maintenance requirements.

3.2 Criteria and Sub-criteria Determining the Model

In this study, we have followed the line of development of the BQA model in Ho (2000). The framework itself can be separated into two main elements, namely the "user element" and the "technical element" (Lawrence and Pande, 2000). Each element is sub-divided into three modules. Building management, access and circulation, and amenities are "user elements," whereas presentation, building services, and structural and system defects are "technical elements". Users may have different perceptions when compared to the experts. Separate analyses on users' and experts' opinions are provided in this study.

The model is composed of a 6 x 3 criteria matrix. The six main criteria, as shown in Figure 7.1, are (1) presentation, (2) maintenance and management, (3) structural and system defects, (4) services, (5) access and circulation and (6) amenities. Within each of the main criteria, there are three sub-criteria relevant to building specifications of residential buildings in Hong Kong.

Figure 7.1   Criteria and Sub-criteria of the VAI

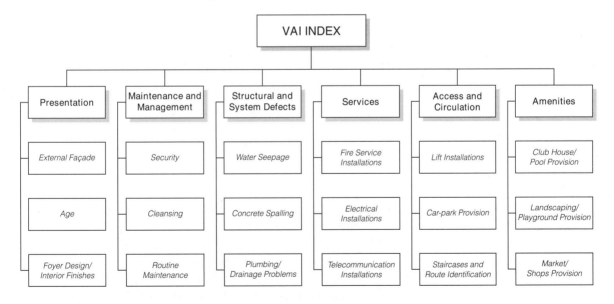

Apart from choosing elements that are relevant to the building type, Soebarto and Williamson (2001) suggested that an effective and useful building quality assessment tool should be able to reflect the relative importance of the various criteria. In our study, instead of assigning arbitrary weights to the criteria, the weights are determined by means of an analysis of the views of both users and experts.

There is a tendency for people to believe that opinions from experts are more reliable and precise than those of users. Therefore, in most cases, experts are chosen to be interviewed, leaving the opinions of users or occupants of buildings aside. Baird, *et al.* (1996) argued that some technical issues regarding building performance are better addressed by an expert. The opinions of the 'true' experts (i.e. the occupants of buildings) may reflect a better understanding of the building's performance. In order to strike a balance between the views of occupants and experts, we have selected a wide range of interviewees, including tenants and owners, as well as experts. By doing this, we hoped to find out if there was a difference between the weights applied to the different criteria by the occupants and the experts.

4 Methodology

The Non-structural Fuzzy Decision Support System (NSFDSS)

The Non-structural Fuzzy Decision Support System (NSFDSS) is a system developed to prioritize complex multi-criteria problems (Tam, *et al.*, 2002 and Chen, 1998). Figure 7.2 shows the flowchart of the NSFDSS process, which is based on three principles:

1. Decomposition: the breakdown of the problem into different levels of independent elements in order to work from the most general level (that is, the goal, and in this study the VAI) to the more specific levels (sub-criteria).
2. Comparative judgment: the construction of the pair-wise comparisons of the relative importance of elements within a given level of quality categories.
3. Synthesis of priorities: the global priority of the elements is given by multiplying local priorities by the priority of the element's corresponding criterion on the level above, and then repeating the procedure all the way to the bottom level.

Figure 7.2   *Flow Chart of the Non-structural Fuzzy Decision Support System (NSFDSS)*

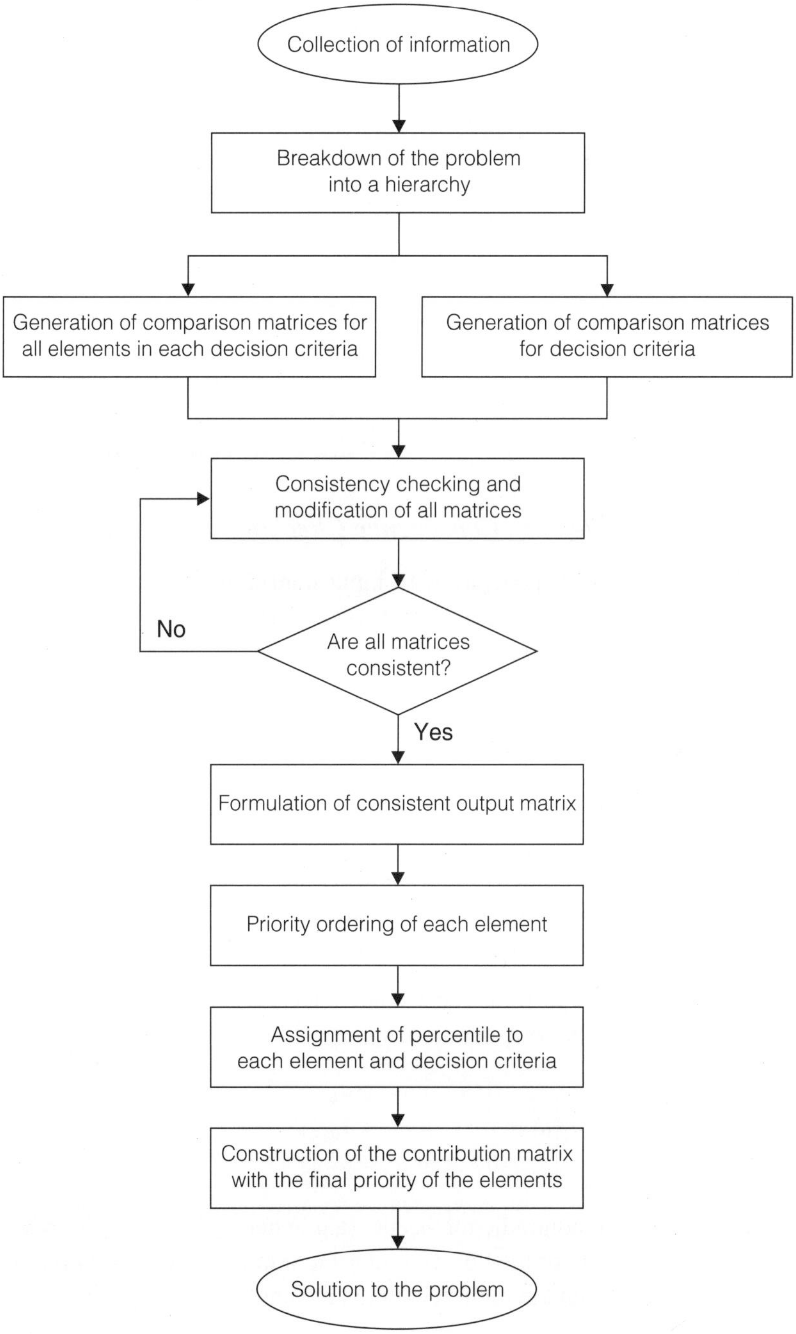

```
                    ┌─────────────────────────┐
                   ( Collection of information )
                    └─────────────────────────┘
                               │
                    ┌─────────────────────────┐
                    │ Breakdown of the problem │
                    │     into a hierarchy     │
                    └─────────────────────────┘
                               │
               ┌───────────────┴───────────────┐
    ┌──────────────────────────┐   ┌──────────────────────────┐
    │ Generation of comparison │   │ Generation of comparison │
    │ matrices for all elements│   │   matrices for decision  │
    │  in each decision criteria│  │         criteria         │
    └──────────────────────────┘   └──────────────────────────┘
               │
    ┌──────────────────────────────┐
    │  Consistency checking and    │◄──────┐
    │  modification of all matrices │      │
    └──────────────────────────────┘      │
               │                          │
            ◇ Are all matrices ◇   No     │
            ◇   consistent?    ◇──────────┘
               │ Yes
    ┌──────────────────────────────┐
    │ Formulation of consistent    │
    │      output matrix           │
    └──────────────────────────────┘
               │
    ┌──────────────────────────────┐
    │ Priority ordering of each    │
    │         element              │
    └──────────────────────────────┘
               │
    ┌──────────────────────────────┐
    │  Assignment of percentile to │
    │ each element and decision    │
    │          criteria            │
    └──────────────────────────────┘
               │
    ┌──────────────────────────────┐
    │ Construction of the          │
    │ contribution matrix with the │
    │ final priority of the elements│
    └──────────────────────────────┘
               │
        ( Solution to the problem )
```

Step 1:   Pair-wise Comparison

In the first step, an input matrix (Figure 7.3 shows an example) is generated by pair-wise comparison between any two elements. There are three scales in the pair-wise comparison: 0, 0.5, and 1, where 0 means that the first element is less important than the second, 0.5 means that they weigh the same, and 1 means that the first element is more important than the second.

Figure 7.3   Example of an Input Matrix

	1	2	3	4
1	0.5	1	0	0
2		0.5	0	1
3			0.5	1
4				0.5

Step 2:   Consistency Checking

After generating an input matrix, consistency checking of the matrix is required. The matrix of a pair-wise comparison of the corresponding element is:

$$A \begin{bmatrix} a_{11} & a_{12} & \cdots & a_{1n} \\ a_{21} & a_{22} & \cdots & a_{2n} \\ \cdot & \cdot & \cdot & \cdot \\ a_{n1} & a_{n2} & \cdots & a_{nn} \end{bmatrix} = (a_{kl}) \qquad k = 1, 2, \ldots, \text{n}; \quad l = 1, 2, \ldots, \text{n} \qquad (1)$$

In the above matrix, a_{kl} is the logical indicator of pair-wise comparison with elements "k" and "l", and n is the number of elements to be considered.

To perform consistency checking, the output matrix is derived from the following conditions:

$$\begin{cases} a_{kl} = 0 & \text{if } a_{hk} > a_{hl} \\ a_{kl} = 1 & \text{if } a_{hk} < a_{hl} \\ a_{kl} = 0.5 & \text{if } a_{hk} = a_{hl} \end{cases} \qquad (2)$$

If contradiction occurs (e.g. when $a_{14} > a_{15}$, but $a_{24} < a_{25}$), then a_{kl} may be determined by counting the number of each relative scale, and the scale with the highest count is the answer to a_{kl}.

Since the matrix of pair-wise comparison is a square matrix and the input matrix only consists of the upper triangle, the lower triangle is obtained by subtracting the upper triangle from 1, and the combination of the two triangles is the output matrix.

Step 3:   *Priority Ordering and the Assignment of Priority Scores to Element*

The aim of this step is to give priority ordering to the elements. First, we sum up the values in each row and rearrange in descending order with respect to criterion C_n. Based on the priority order, experts can assign a percentile to each element. With the top element (the element with the highest score) as 100%, the remaining elements are compared to it one by one to distinguish the importance between them. The percentile should thus decrease as the comparison proceeds. Each percentile is assigned a score, sj, in the range of [1, 0.5], with 1 representing "same importance" and 0.5 representing "not important," as shown in Table 7.1. Then the score will be converted into a priority score, rj, in the range of [1,0] by applying fuzzy set theory through the following equation:

$$r_j = \frac{1 - s_j}{s_j}, 0.5 \le s_j \le 1 \tag{3}$$

Table 7.1   Priority Score

Percentile (%)	s_j	r_j
100	0.500	1.000
95	0.525	0.905
90	0.550	0.828
85	0.575	0.739
80	0.600	0.667
75	0.625	0.600
70	0.650	0.538
65	0.675	0.481
60	0.700	0.429
55	0.725	0.379
50	0.750	0.333
45	0.775	0.290
40	0.800	0.250
35	0.825	0.212
30	0.850	0.176
25	0.875	0.143
20	0.900	0.111
15	0.925	0.081
10	0.950	0.053
5	0.975	0.026
0	1.000	0.000

Step 4: Determination of the Results

After gathering the weights of each criterion and sub-criterion, a contribution matrix (CM) can be constructed to show the overall ranking of each element. The general ranking is obtained by multiplying the weight of each element with the weight of the respective decision criteria, that is: $CM_{ij} = w_i * r_{ij}$ where: CM_{ij} = contribution of each element in the problem; w_i = the weight of decision criterion "i"; and r_{ij} = the weight of element "j" for decision criterion "i".

For more details of the NSFDSS, readers may refer to Tam, *et al.* (2002) and Chen (1998).

5 | Questionnaire Survey

We conducted a questionnaire survey with 61 face-to-face interviews. Each respondent was asked to assess the importance of the criteria by means of the NSFDSS. For instance, if a respondent believed that the seepage problem affects building quality more than the defects due to concrete spalling, then he or she would place a greater weight on the seepage symptom.

The respondents are divided into two groups: experts and non-experts. The experts are mainly building surveyors, as they are trained and experienced in carrying out building condition surveys. In addition to discerning the difference of opinions between experts and non-experts, male and female respondents were analyzed separately as well. Table 7.2 shows the distribution of respondents, which reveals that the distribution was balanced between expert and non-expert, male and female respondents.

Table 7.2   Distribution of Respondents

Category	Male	Female	Expert	Non-expert
No. of Respondents	35	26	33	28
Total	61		61	

6 **Results**

Figure 7.4 shows the overall weights of the 18 sub-criteria in descending order. It shows that the three sub-criteria under "structural and system defects" are of top priority. They are, on average, almost twice as important as the least important ones. The least important criteria are those under the heading of "Amenities".

Figure 7.4   Weights of the 18 Sub-criteria

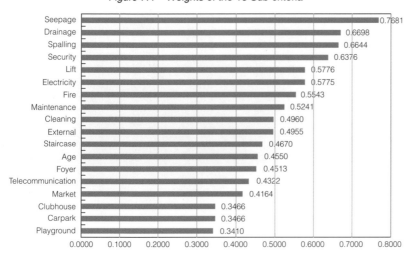

This shows that in general, people place the highest value on the structural and system defects of a building, followed by building services, management, access, and circulation, while the least important criterion is the provision of amenities. Obviously, the structural and system defects of a building greatly affect the efficiency of a building. The problem of concrete spalling directly threatens people's lives, and in turn adversely affects the value of a building. The provision of amenities ranks the lowest. Such findings are plausible, since the provision of amenities is regarded as a supplement to the actual function of a building, rather than being part of its own value.

It is also interesting to find out that the age of a building, a well-recognized critical element in building quality assessment, is not as important as structural and system defects, building services, and building management. We infer that since age has a high correlation with the quality of a building in terms of structural and system defects and building services, it is taken as a proxy in the determination of a building's quality. However, when those criteria of structural and system defects and the provision of building services are directly and separately taken into account, the building age itself becomes much less important.

Table 7.3   Weightings by Different Groupings of Interviewees

Element	Experts	Non-experts	Male	Female	Overall
External	0.5573	0.4151	0.5007	0.4789	0.4955
Age	0.4873	0.4130	0.4526	0.4625	0.4550
Foyer	0.5186	0.4151	0.4582	0.4625	0.4513
Security	0.5523	0.7684	0.6297	0.6625	0.6376
Cleaning	0.4596	0.5433	0.4679	0.5854	0.4960
Maintenance	0.4594	0.6082	0.5106	0.5671	0.5241
Seepage	0.7961	0.7316	0.7763	0.7421	0.7681
Spalling	0.6251	0.7165	0.6706	0.6448	0.6644
Drainage	0.6637	0.7896	0.6847	0.7479	0.6698
Fire	0.4702	0.6636	0.5510	0.5648	0.5543
Electricity	0.5046	0.6723	0.5605	0.6317	0.5775
Telecommunications	0.3609	0.7468	0.4025	0.5268	0.4322
Lift	0.5349	0.6331	0.5697	0.6025	0.5776
Staircases	0.3937	0.5404	0.4492	0.5141	0.4670
Car-park	0.3108	0.3932	0.3426	0.3597	0.3466
Clubhouse	0.3787	0.4081	0.3781	0.4543	0.3466
Playground	0.3597	0.3167	0.3358	0.3575	0.3410
Market	0.4438	0.3822	0.3942	0.4851	0.4164

Figure 7.5   Experts' Weighting

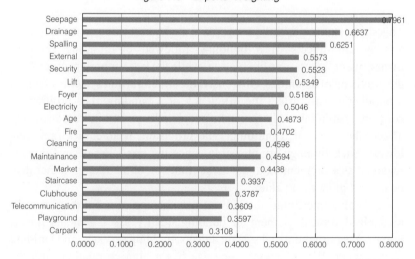

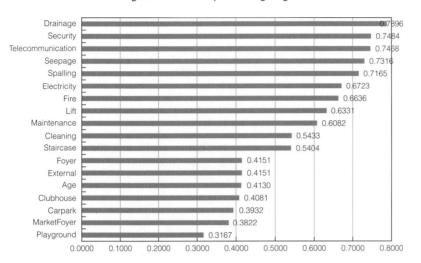

Figure 7.6   Non-experts' Weighting

Table 7.3 shows the weightings assigned by different groups, including experts, non-experts, males, and females. Figures 7.5 and 7.6 depict the rankings in descending order for expert and non-expert weightings respectively. From the results, we see that structural and system defects, such as seepage and concrete spalling defects, are still regarded as the most important criteria by the interviewees under different groupings.

However, non-parametric statistical tests (not shown) reveal that opinions are significantly different in some of the other criteria. For example, these tests show that male and female respondents have significantly different weightings towards all criteria, except for cleansing, seepage, and lift provision. These tests show that the views of experts and non-experts are also notably different, except for seepage and lift provision.

7 Conclusion

Having conducted the questionnaire survey, the results showed that the age of a building, the well-recognized criterion in building value determination, is not as important as some other criteria. The most important criteria are structural and system defects and building services of a building. It was also found that different groups of interviewees place different weightings on the criteria provided. This implies that if a generalized index is to be constructed, then different weightings among groups of interviewees should be considered; otherwise the result would be biased. This study is conducive to our further research study in constructing a more comprehensive VAI, and to developing other modules in overall building quality assessment.

References

1. Baird, G., Gray, J., Isaacs, N., Kernohan, D. and McIndoe, G. 1996. *Building Evaluation Techniques*. New York: McGraw-Hill.

2. Chen, S. Y. 1998. *Engineering Fuzzy Set Theory and Application*. Beijing, China: State Security Industry Press.

3. Cliff, M. 1996. Building Quality Assessment (BQA) for Offices, *Structural Survey*, 14(2), 22–25.

4. Davis, G., Thatcher, C and Blair, L. 1993a. *Serviceability Tools, Vol. 1, Methods for Setting Occupant Requirements and Rating Buildings*. International Centre for Facilities, Ottawa.

5. Davis, G.; Gray, J. and Sinclair, D. 1993b. *Serviceability Tools, Vol. 2, Scales for Setting Occupant Requirements and Rating Buildings*. International Centre for Facilities, Ottawa.

6. Ho, D. C. W. 2000. *An Analysis of Property-Specific Quality Attributes for Office Buildings*. Unpublished Ph.D. Thesis, The University of Hong Kong, Hong Kong SAR.

7. Ho, D. C. W. 1999. Preferences on Office Quality Attributes: A Sydney CBD Study, *Australian Land Economics Review*, 5(2), 36–42.

8. Issacs, N. Bruhns, H. Gray, J. and Tippett, H. 1994. *Building Quality Assessment — Research, Development and Analysis for Office and Retail Buildings*. Centre for Building Performance Research, Victoria University of Wellington.

9. Kaiser, H. H. 1989. *The Facilities Manager's Reference*. New York: R S Means Co Inc.

10. Lai, L. W. C. & Ho, D. C. W. 2001. Unauthorized Structures in a High-rise High-Density Environment: the Case of Hong Kong, *Property Management*, 19(2), 112–123.

11. Lawrence, D. and Pande, R. 2000. The Facility Audit: A User Oriented Design Paradigm. *Proceedings of CIB W70 Symposium*, Brisbane.

12. Rider Hunt Group. 1991. *Building Quality Assessment*. Sydney, Australia: Rider Hunt Group.

13. So, A. T. P. and Wong, K. C. 2002. On the Quantitative Assessment of Intelligent Buildings, *Facilities*, 20(5/6) 208–216.

14. Soebarto, V. I. and Williamson, T. J. 2001. Multi-criteria assessment of building performance: theory and implementation, *Building and Environment*, 36, 681–690.

15. Tam, C. M., Tong, T. K. L., Chiu, G. C. and Fung, I. W. H. 2002. Non-Structural Fuzzy Decision Support System for Evaluation of Construction Safety Management system. *International Journal of Project Management*, 20, 303–313.

8

A Systematic Framework of Fire Risk Ranking of Existing Buildings

S. M. Lo and Grace W. Y. Cheng

The assessment of the fire safety level of older buildings in Hong Kong — on the basis of current prescriptive requirements — might return a conclusion that their fire safety systems are "sub-standard", and that the fire risk is unacceptably high. However, whether such a conclusion is warranted, thereby triggering immediate remedial action, is open to debate because the rigid prescriptive requirements in the fire codes do not provide a holistic picture of the fire safety level in these buildings. If all of the buildings that were constructed according to previous prescriptive requirements were required to improve at the same time, society would have difficulty shouldering the cost. The prioritization of improvement works for fire protection in buildings is necessary. This chapter describes a systematic evaluation technique, and an example is used to illustrate the specific issues.

Chapter

A Systematic Framework of Fire Risk Ranking of Existing Buildings

S. M. Lo and Grace W. Y. Cheng

1 Introduction

Numerous fire disasters in recent years have shown that fire safety should become a major concern. A large number of high-rise buildings constructed before the 1980s were designed according to old prescriptive building and fire codes. The fire protection measures of these buildings may not match up to current standards, even if all fire safety items have been well maintained. An assessment of the fire safety level of these older buildings — on the basis of current regulations — may lead us to conclude that the fire safety systems of many buildings are sub-standard, and that the fire safety level is unacceptably low. However, whether or not this conclusion is warranted, instant remedial action may not be the right course, as the rigid fire regulations do not provide a holistic picture of the fire safety level in these buildings. Indeed, if all the buildings that have been constructed according to previous regulations are required to improve at the same time, then the cost would be hard to bear. Therefore, a process to evaluate the fire safety level of old buildings should be established. The ranking established can serve as a reference for decision-makers, such as the building owners, to judge whether improvement works should be carried out.

2 Previous Studies

Fire safety assessment can be carried out by various approaches (NFPA101A, 1995; Pate-Cornell, 1995; Frantzich, 1998; Chow *et al.*, 1999, Chow and Lui, 2001, Chow, 2002; Lui and Chow, 2000). A deterministic engineering approach (BSI, DD240: 1997) can provide a sound evaluation on the basis of fire dynamics, fire chemistry, evacuation principles and so on. However, the engineering approach may be difficult to apply to a complex building and wherever a holistic assessment is demanded. In these cases, an evaluation on the basis of an auditing approach may be useful. This approach can serve as a decision-making tool to rank the priority of improvement works.

Watts (1997) reminds us that:

> "Fire risk ranking is a method of fire risk assessment. It constitutes various processes of analyzinanalysing and scoring risk parameters to produce a rapid and simple estimate of relative fire risk. Values are assigned to the parameters (passive and active fire safety features) based on professional judgement and past experience and then aggregated by some arithmetic function to arrive at a safety index. The safety index can be compared to other similar assessments to rank the fire risk."

Such methods originated with insurance rating schedules and have been developed in various applications (Dow Chemicals, 1993, 1994, 1966; International Atomic Energy, 1993, Nelson *et al.*, 1984). For example, in the USA, a fire safety evaluation system (FSES) has been developed on the basis of fire risk ranking (NFPA 101, 1994, National Fire Protection Association 101A, 1995). The FSES provides a multi-attribute approach to determine equivalencies to the National Fire Protection Association 101 Life Safety Code (1991) for certain occupancies. The Central Office Fire Risk Assessment (COFRA) methodology has been developed for telecommunications central offices (Parks *et al.*, 1998). A fire safety evaluation system for National Park's service overnight accommodation by Nelson *et al.*, (1984) [10] has been developed for assessing existing buildings (Nelson, 1984). In Europe, Gretener (1973, 1980) developed a fire insurance evaluation method. Marchant (1982) developed the Edinburgh Model by using the Dephi Approach and Shield (1986) developed a fire safety evaluation method by using an analytical hierarchy process (AHP). Lo (1999) has developed an evaluation method on the basis of fuzzy set theory (Zadeh, 1965, 1973). This article adopts the latter model to illustrate the ranking process.

3   The Approach

The evaluation involves various eight stages:

1.   Identify the objectives of the evaluation
2.   Formulate suitable methods for the evaluation
3.   Establish the attributes (components) of the evaluation framework
4.   Determine the synthetic process of the evaluation
5.   Establish the measurement guidance for the fire safety attributes
6.   Perform an inspection (measure the scores of the attributes)
7.   Compute the results
8.   Produce a ranking for reference purposes

Items 1 and 2 have been discussed above, so the following will concentrate on items 3 to 8.

3.1   Establishment of Attributes

Identification of Attributes

Fire safety is a complex system with large number of attributes that may affect it. The overall level of safety will be determined by the attributes affecting fire safety. Although the attributes are numerous, a relatively small number of factors account for most of the problems (Watts, 1995, 1997b) and it is possible to reduce the large number of parameters (attributes) to an appropriate number which can be handled with appropriate effort. For example, the NFPA101A addresses 14 general areas affecting building life fire safety of buildings (Watts, 1997a).

In order to facilitate the process, we need to identify the attributes and to assess the weightings of each attribute. Figure 8.1 briefly outlines the way to set up the attributes.

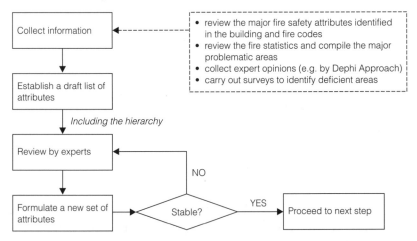

Figure 8.1 *Establishment of Fire Safety Attributes*

3.2 Establishment of Weightings

In order to implement the assessment model, it may be required to establish the relative the importance of an attribute may be required to identify, in particular if it is required to specify the differential importance of the attributes. As the attributes that need to be handled are numerous, a hierarchy of different levels can be formed for comparison. Moreover, some of the attributes are difficult to quantify. Evaluation by experts is one of the easier ways to determine the relative importance of each attribute. It has proposed in various studies (Shield, 1986; Watts, 1995; Lo, 1999) that the use of Saaty's analytical hierarchy process is a useful method for determining the weightings of the attributes. However, the use of other methods, such as direct point allocation, multiple regression models, explicit trade-offs (Keeney and Raiffa, 1993), equal/unit weighting and other optimization methods can also be considered. A method termed EB-FSRS (Chow, 2002), in which weightings may not be required for the evaluation, has also been proposed.

3.3 The Synthetic Process

The overall level of safety will be determined by the parameters affecting fire safety (Lo, 1999).

$$\text{thus} \qquad FSL = f(SL_E, SL_S, SL_C, SL_A, \ldots\ldots) \qquad (1)$$

where   FSL = total fire safety level
SL$_E$ = safety level with respect to exit routes
SL$_S$ = safety level with respect to smoke control, etc.

FSL is a function of a list of fire safety attributes and such attributes are not directly measurable. This is especially true for existing building where only limited information is readily available.

The operation of equation (1) should be simple from the practical point of view and flexible from the management point of view. In fire safety systems, only incomplete data and information are available. The evaluations of fire safety attributes which are carried out by experienced investigators[1] (surveyors) are predominantly subjective and imprecise. It has also been pointed out in (Watts, 1995) that most fire safety problems are not clear cut and fuzzy logic approach may be a way to provide a reasonable answer. Such an approach is not as complicated as a probabilistic approach and is considered more practicable (Lo, 1999). In this approach every item to be evaluated evaluation work identified during an inspection by suitable surveyors is measured and a score (in linguistic terms or in a fuzzy number) will be given in respect of each criterion determined.

A brief description of the evaluation process is given below:

Let m criteria be used in the evaluation process for an occupancy j in that a set of attributes C_j can be expressed as:

$$C_j = \{c_{j1}, c_{j2}, \ldots\ldots\ldots, c_{jm}\} \tag{2}$$

We denote an evaluation set E_j with n evaluation levels representing the FSL and can be expressed as:

$$E_j = \{e_{j1}, e_{j2}, \ldots\ldots\ldots, e_{jn}\} \tag{3}$$

Now, we define a set of evaluations of individual attributes:

$$R^* = \{\frac{r_1{}^*}{c_1}, \frac{r_2{}^*}{c_2}, \ldots\ldots \frac{r_m{}^*}{c_m}\} \tag{4}$$

where $r_k{}^*$ is the score given by the surveyor.

R^* can be expressed as an m-row evaluation matrix:

$$R^* = \begin{matrix} MoE \\ MoA \\ : \\ : \end{matrix} \begin{bmatrix} r_1{}^* \\ r_2{}^* \\ : \\ : \end{bmatrix} \tag{5}$$

Suppose there exists a mapping such that:

$$\phi : R^* \to R \tag{6}$$

where ϕ is called the membership function of R^* which can be represented by a triangular fuzzy number.

Then,

$$R = \begin{array}{c} MoE \\ MoA \\ : \\ : \end{array} \begin{bmatrix} \varphi(r_1{}^*) \\ \varphi(r_2{}^*) \\ : \\ : \end{bmatrix} \tag{7}$$

For m rows of criteria and n columns of levels, R can be written as:

$$R = \begin{bmatrix} r_{11} & r_{12} & .... & r_{1n} \\ : & : & .... & : \\ : & : & .... & : \\ r_{m1} & & .... & r_{mn} \end{bmatrix} \tag{8}$$

Now, let A = the weighting of the attributes in C

$$W = \{w_1, w_2, ... w_m\} \tag{9}$$

Then, the evaluation B can be given by:

$$B = W \circ R \tag{10}$$

B is the direct product: $W \times R = \{w, r \mid w \in W, r \in R\}$, and is characterized by a membership function in $[0, 1]$.

The composition of equation (11) can be handled by:

$$: b_j = \min\{1, \sum_{i=1}^{m} w_i r_{ij}) \qquad \forall \quad b_j \in B \tag{11}$$

This approach is suitable for the problems where many attributes are considered and the difference of the weighting for each attribute is not great. Accordingly, this model is suitable for fire safety ranking.

The composition result, B, given by (11) can provide the relative information for ranking purpose.

4 Discussion

A notional trial of the proposed evaluation framework for fire safety ranking can be carried out. As some of the residential buildings in Hong Kong are of similar form and occupancy, they can serve as an example to illustrate the evaluation. We can simplify this exercise by selecting typical residential buildings as the target buildings because they are of a similar management level and occupancy pattern. The major difference among these buildings is their orientation and design and their completion date.

4.1 Identification of Experts

Building professionals in Hong Kong, in particular building surveyors, have adequate experience in carrying out maintenance works in these buildings. Their fire safety criteria and attributes can therefore be easily identified with the expert opinion of these professionals. A group of building surveyors can be asked to assist in the identification of fire safety criteria and attributes with respect to a typical residential building.

4.2 Formulation of Attributes

In order to simplify the process of a trial exercise, the hierarchy can be limited to three levels and the formulation may be carried out in two-stages. In the first stage, the experts identified are asked to identify the major fire safety criteria (see Figure 8.2) for typical residential buildings. The responses from these experts form a sheaf. The percentage of identification[2] of each criterion are is then sent back to each individual expert. Each expert is then requested to provide a new list of criteria in order to group some of them. The process is repeated until the percentage of identification of each criterion becomes sufficiently stable (Figure 8.1 refers). A list of fire safety criteria is formulated by discussing them individually with each expert. In the second stage, the experts are asked to identify the major attributes of each fire safety criterion. A process similar to that in the first stage is adopted. When the percentage of identification of each attribute becomes stable, a hierarchy of fire safety criteria and attributes can be formed (Figure 8.2 refers).

Figure 8.2   The Hierarchy of Attributes Identified

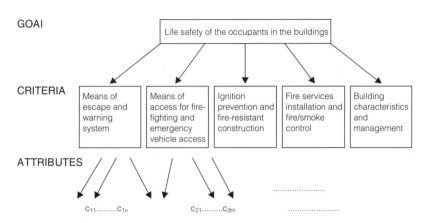

GOAl

Life safety of the occupants in the buildings

CRITERIA

| Means of escape and warning system | Means of access for fire-fighting and emergency vehicle access | Ignition prevention and fire-resistant construction | Fire services installation and fire/smoke control | Building characteristics and management |

ATTRIBUTES

$C_{11}.........C_{1n}$ $C_{21}.........C_{2m}$

Remarks: i) attributes identified are listed in Table 1;
            ii) to facilitate the experts to identify the attributes, the wording and grouping are in
                accordance with the fire and building codes used in Hong Kong.

4.3   Formulation of Weightings

The relative importance of the criteria and attributes can be determined on the basis of Saatty's analytical hierarchy process (AHP) by using the computer package, EXPERT CHOICE. The experts were asked to give verbally the relative importance of the criteria and attributes by pairwise comparison. The weightings for residential buildings have been enformulated and are shown in Table 8.1 (Lo, 2000).

4.4   The Survey

The survey items are the measurable building and space attributes identified in Table 8.1. The surveyors will be required to give grades (in linguistic terms, such as "very safe", "safe", and "moderate") to the attributes for individual buildings. In implementing the survey, brief explanations of the attributes should be given to the surveyors. They should be required to grades each item according to the guidelines and their professional experience.

Table 8.1   *Weightings of Fire Safety Criteria and Attributes*
for Residential Buildings (Lo, 2000)

Fire Safety Criteria/Fire Safety Attributes		Overall Weighting
Means of Escape	Number of Exits	0.046
	Travel/Direct Distance	0.061
	Width of Exit Route	0.040
	Exit Signs	0.067
	Emergency Lighting	0.078
	Refuge Floor	0.023
Fire Resistant Construction	Compartmentation	0.034
	Fire Rated Door	0.047
	FR Walls/Floors	0.029
	External Flame/Smoke Spread	0.042
	Internal Flame/Smoke Spread	0.034
Means of Access for Fire Fighting and Rescue	FF and R stair	0.023
	Firemen's Lift	0.017
	Smoke Vent	0.015
	Emergency Vehicular Access	0.020
	Ventilated Lobby	0.018
Fire Services Installations	Fire Hydrant	0.016
	Fire Alarm	0.027
	Hose Reel	0.033
	Fire Blanket	0.016
	Auto-detection	0.028
	Sprinkler System	0.071
	Portable Fire Extinguisher	0.025
Building Management and Maintenance	Management Level	0.072
	Staff Training	0.065
	Maintenance Level	0.045

Remarks: The weightings were obtained from the expert panels comprising building surveyors and
building services engineers.

5   Limitations

Fire risk ranking methods, like other analytical techniques, have their
limitations and should not be used uncritically. The purpose of fire risk
ranking is to provide a useful aid to decision-making. Decisions on the
ranking of a building are surrounded by crude and subjective assessments.
Their credibility and applicability are considered in respect of the 10
criteria indicated by Watts (1997b).

1.   The development and implementation of the method should be
documented according to pre-determined standard procedures.

The structure of the model, including the assumptions and constraints, should be clearly stated and documented in order to allow users to understand the rationale behind it. The information should include the procedures and criteria in selecting the members of the expert panel, the method used in formulating the fire safety attributes, the technique in establishing the weightings of the attributes, the criteria applied in giving the grades of the attributes and the synthesis process resulting in the final ranking of a building.

2. The attributes selected should be as comprehensive as possible. The attributes identified by the experts can be numerous. A reduced set of the attributes may cause some fire safety issues be ignored. Some may comment that the three-level hierarchy should be extended to include sub-attributes that elaborate the meaning of each attribute and serve as a checklist for inspection. This, however, will make the checklist too complicated to use.

3. Attributes should represent the most frequent fire scenarios. Too many attributes may cause problems in itswith implementation. The attributes that are most significant, whether statistically or those that have been identified as such by experienced judges should be adopted. Accordingly, we consider that different sets of attributes for different types of buildings should be adopted.

4. Operational definitions should be provided for the attributes. If the technique is to be used by more than a single surveyor, it is necessary to ensure that the intent of the attributes is clearly communicated. Guidelines should be given to the surveyors for the evaluation process. Although they may use their professional judgement to evaluate the attributes, a clear basis for judging each item should be used to ensure consistency. Furthermore, appropriate training on the fire safety survey will assist the surveyors to provide a standard evaluation of the attributes.

5. The subjective values for evaluation should be elicited systematically. Nevertheless, a user-friendly computer program should be developed for complicated calculations. It is preferable to ask for input in linguistic terms.

6. The values of the attributes should be maintainable. Technological advancement and the changing safety level

expected in society will make it necessary to review the set of attributes from time to time. A continuous monitoring system and a two-year review period may be considered and may be different for different organizations or societies.

7. The interaction of attributes should be treated consistently. It was believed that some attributes are indeed interactive with other attributes. Clarification in respect of this issue should be given to the surveyors.

8. Assumptions should be stated clearly. Watts (1997b) pointed out that most fire ranking methods are assumed to behave linearly ranked. The use of fuzzy arithmetic allows us to include the transformation of imprecise evaluation by a predetermined function. The usefulness of a five, seven or nine grade system for judging should be carefully studied. A five-grade system may cause difficulties in distinguishing the risk ranking between similar buildings.

9. Fire risk should be described by a single indicator. The evaluation should be based only on technical judgements. Other factors, such as socio-economic aspects, should not be included in establishing fire safety grades. They should be considered only at the management level in order to set the priorities of improvement works. Another ranking system may be needed to assist decision-making at management level.

10. The results should be verified. The use of sophisticate mathematical and probabilistic models could provide a comparatively more accurate result. However, it would not be easy to apply for a massive evaluation of complex existing buildings. The proposed method could be a practical way to solve the problem initially and its credibility would depend upon the user seeing how well it worked. In general, the question of credibility may be viewed as consisting of two components, namely internal validity, which concerns logic, reasoning and analysis, and external validity, which asks whether the premises and assumptions accord with empirical reality. The internal validity of the proposed method can be performed by verifying the logic and reasoning. The participation of various experts in developing the system would improve the internal validity of the system. Limited statistical data combined with building

complexity may make empirical verification difficult. How much weight one places on internal, versus external, validity is ultimately a matter of psychological and philosophical preference. Alternatively, we may consider developing an evaluation method in which weighting is not necessary (Chow, 2002).

6   Concluding Remarks

Buildings will deteriorate during their life span. In respect of the fire safety standard in a building, a ranking method may be adopted to determine the fire safety level and from which to prioritize actions for improvement. Fire risk ranking is not a new system for use in fire safety assessment. It has been adopted to demonstrate they ways in which it is equivalent to prescriptive requirements (NFPA 101A, 1998). Developing fire risk ranking into a multi-attribute fire safety evaluation system can provide a standard assessment tool for the professionals and the regulatory bodies to follow. In view of the imprecise and vague nature of some of the values of the attributes, the evaluation system can be established on the basis of fuzzy set approach. In the Hong Kong situation, both the buildings and the environmental conditions are complex. Fire safety models based on fundamental physical, chemical, thermodynamic and psychological principles may be difficult to carry out efficiently for massive evaluation. A simple evaluation tool may assist the decision-makers, such as the regulatory bodies as well as the owners of the buildings to set priorities for action, priorities that are also related to the financial and political aspects of safety. In order to develop a creditable ranking system, a careful study of the identification of fire safety parameters (attributes), synthetic methods and inspection and measurement guidelines should be carried out. In addition, a computer package may be developed for the implementation of the computation process. At all events, a monitoring system should be established to review the implementation of the system and provide refinements to it. This paper outlines the framework of the system: further studies are required to provide details on how to set it up.

Notes

1. The surveyor responsible for the inspection should have adequate training in building surveying technique and have adequate knowledge of fire safety design in buildings.

2. The percentage of identification of an attribute j is defined as:
   $\%I_j$ = (number of I_j)/(number of experts)

References

1. Chow, W. K. 2002. Proposed Fire Safety Ranking System EB-FSRS for Existing High-Rise Non-Residential Buildings in Hong Kong. *ASCE Journal of Architectural Engineering*, 8, 116–124.

2. Chow, W. K., Wong, L. T. and Kwan, E. C. Y. 1999. A Proposed Fire Safety Ranking System for Old High-Rise Buildings in the Hong Kong Special Administrative Region. *Fire and Materials*, 23, 1, 27–31.

3. Chow, W. K. and Lui, G. C. H. 2001. A Fire Safety Ranking System for Karaoke Establishments in Hong Kong. *Journal of Fire Science*, 19,2, 106–120.

4. Dow Chemical Ltd. 1993. *Dow's Chemical Exposure Index Guide*. American Institute of Chemical Engineers, New York.

5. _____. 1994. *Dow's Fire and Explosion Index Hazard Classification Guide. 7th edn*. American Institute of Chemical Engineers, New York.

6. _____. 1966. Process Safety Manual. *Chemical Engineering Progress*. 62, 6.,

7. Gretener, M. 1973. *Evaluation of Fire Hazard and Determining Protective Measures*, Association of Cantonal Institutions for Fire Insurance (VKF) and Fire Prevention Service for Industry and Trade (BVD), Zurich.

8. Gretener, M. 1980. *Fire Risk Evaluation*. Association of Cantonal Institutions for Fire Insurance (VKF), Society of Engineers and Architects (SIA) and Fire Prevention Service for Industry and Trade (BVD), Zurich.

9. International Atomic Energy Agency. 1993. *Manual for the Classification and Prioritisation of Risks Due to Major Accidents in Process and Related Industries*. International Atomic Energy Agency, Vienna.

10. Keeney, R. L. and Raiffa, H. 1993. *Decision with Multiple Objectives: Preferences and Value Trade-Offs*. Cambridge: Cambridge University Press.

11. Lo, S. M. 1999. A Fire Safety Assessment System for Existing Buildings. *Fire Technology*, .

12. Lo, S. M., Lam, K. C. and Yuen, K. K. 2000. Views of Building Surveyors and Building Services Engineers on Priority Setting of Fire Safety Attributes for Building Maintenance. *Facilities*, 18, 13/14, 513–523.

13. Lui, G. C. H. and Chow, W. K. 2000. A Demonstration on Working Out Fire Safety Management Schemes for Existing Karaoke Establishments in Hong Kong. *International Journal on Engineering Performance-Based Fire Codes*, 2, 3, 104–123.

14. Marchant, E. W. 1982. *Fire Safety Evaluation (Points) Scheme for Patient Areas within Hospitals*. Report, Department of Fire Safety Engineering, University of Edinburgh.

15. National Fire Protection Association. 1994. NFPA101, *Life Safety Code*. Quincy, MA.

16. _____. 1995. NFPA101A, *Guide on Alternative Approaches to Life Safety*. Quincy, MA.

17. Nelson, H. E., Shibe, A. J., Levin, B. M., Thorne, S. D. and Cooper, L. Y. 1984. *Fire Safety Evaluation System for National Park Service Overnight Accommodations*. Gaithersburg, MD: National Bureau of Standards, NBSIR 84–2896.

18. Parks, L. L., Kushler, B. D., Serapighlia, M. J., McKenna, L. A. Jr., Budnick, E. K. and Watts, J. M. 1998. Fire Risk Assessment for Telecommunication Central Offices. *Fire Technology*, 34, 2, 156–176.

19. Shields, T. J. and Silcock, G. W. 1986. Fire Safety Evaluation of Dwellings. *Fire Safety Journal*, 10, 29–36.

20. Watts, J. M. 1995a. Fuzzy Fire Safety, *Fire Technology*. 193–194.

21. _____. 1995b. Fire Risk Ranking. In P. J. DiNenno *et al.* (eds) *SFPE Handbook of Fire Protection Engineering*. National Fire Protection Association, Quincy, MA.

22. _____. 1997a. Analysis of the NFPA Fire Safety Evaluation System for Business Occupancies. *Fire Technology*. 33, 3.

23. _____. 1997b. Fire Risk Assessment Using Multi-Attribute Evaluation, 679–690. In Y. Hasemi, (ed.) *Fire Safety Science: Proceedings of the Fifth International Symposium*, Tsukuba:

24. Zadeh, L. A. 1965. Fuzzy Set, *Information and Control*, 8, . 338–353.

25. _____. 1973. Outline of New Approaches to Analysis of Complex Systems and Decision Processesd. *IEEE Transactions on System, Man and Cybernetics*, SMC-3, 3, 1, 28–44.

9

Comparing the Deterministic Fire Engineering Approach and the Fire Safety Ranking Approach

S. M. Lo, Richard K. K. Yuen and Grace W. Y. Cheng

There are numerous methods for evaluating the fire safety level of old buildings. A deterministic engineering approach can provide a sound evaluation on the basis of fire dynamics, fire chemistry, evacuation principles, etc. However, such an approach may be difficult to apply to a complex building and where a holistic assessment is needed. Thus, various semi-quantitative evaluation methods based on the auditing approach have been proposed. The auditing approach is especially useful when a massive evaluation task is needed to rank the fire safety level of numerous old buildings for decision makers to set priorities for improvement works. This chapter uses a case study to compare evaluations that have used the deterministic fire engineering approach and those that have used the fire safety ranking approach.

Chapter

9

Comparing the Deterministic Fire Engineering Approach and the Fire Safety Ranking Approach

S. M. Lo, Richard K. K. Yuen and Grace W. Y. Cheng

1 Introduction

The big fires that have recently happened in Hong Kong (the Garley Building Fire, 1996, with 40 deaths; the Top-One Karaoke Fire, 1997, with 15 deaths) and China (the Luoyang Dongdu Commercial Building, 2002, with 309 deaths) have led the people in China (including Hong Kong) to express grave concern about the fire safety level of buildings. Both the people and the government are particularly concerned about the design and management of buildings in respect of fire safety. In many old buildings, the methods of fire protection are designed in accordance with the old prescriptive building and fire codes. If they are judged on the basis of current regulations, they may well be considered sub-standard. However, this may not in itself be a danger to the occupants of these buildings. We need to consider the fire safety level of a building from a holistic point of view to provide grounds for making careful decisions on whether immediate improvement action should be taken or not.

2 Fire Safety Assessment

There are many approaches to evaluate the fire safety level of a building. In general, the following methods are widely used (Magnusson, 1998):

i) A deterministic fire engineering approach. A methodology, based on physical relationships derived from scientific theories and empirical results, that a given set of initial conditions will always produce the same outcome (BSI, DD240: 1997). Safety criteria can be established with reference to, among others:

    a) The evacuation time for occupants to leave the building.

    b) The critical time to tenability limit.

On the basis of these criteria, the safety of the occupants is assessed, for example, on whether they can be safely evacuated from the hazardous zone before the zone becomes untenable.

ii) A fire risk assessment. This may show that the likelihood of a given event, such as injury, death or a large loss of life, occurring is within an acceptable risk limit (so that, for example, an individual risk level at home at, say 1.5×10^{-5} deaths per year, is acceptable). This probabilistic approach involves more sophisticated mathematical techniques including computer simulation, linear regression and stochastic modelling. It is similar to the deterministic evaluation of an event tree (Magnusson, 1998).

iii) Fire safety ranking. A semi-quantitative method, based on multi-attributes evaluation, may involve the processes of analysing and scoring risk factors (attributes) to produce a rapid and simple estimate of the relative fire safety level of a building. This method is similar to a safety auditing approach (Watts, 1995; NFPA 101A, 1995).

The use of sound fire science and engineering in the deterministic approach is usually considered a reasonable approach to evaluating the safety of the occupants of a building in a fire situation. However, when we need to evaluate a set of buildings and compare their safety level, the fire safety ranking approach may provide a rapid and simple solution. As the fire safety ranking approach does not directly determine the development of a fire or the evacuation of people, it may not have the same credibility as the fire science and engineering approach. Nevertheless, a carefully established fire ranking method can provide valuable information for comparison purpose. This article uses a case study to compare the deterministic approach and the fire safety ranking approach.

3 A Case Study

3.1 Background

One of the three staircases of a 12-storey residential building in Hong Kong was in a dilapidated condition and immediate repair works needed to be carried out. In order to facilitate the repair works, the temporary closure of the dilapidated staircase (the middle staircase) as highlighted in Figure 9.1 was proposed. However, closing the staircase means the travel distance to one part of the building will exceed the requirement stipulated in the Means of Escape Code, 1996. Figure 9.1 shows the layout of the building. We need to discover, by using the analytical approach, whether the proposed arrangement for temporarily closing the mid-stair is viable. Having identified the problem, we need to review the acceptance criteria implied in the codes. Such criteria will serve as the reference for the evaluation.

Figure 9.1   Layout Diagram of the Residential Building

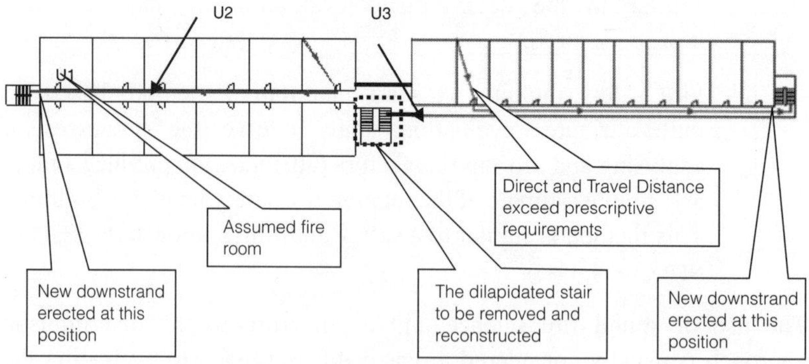

The Code of Practice for the Provision of Means of Escape in Case of Fire 1996 (MoE Code, 96) stipulates that the occupants, in a situation where total evacuation is necessary, should be capable of entering the fully protected zone in 2.5 minutes (paragraph 15 of the commentary in the MoE Code, 96). We need to perform a computational study on the escape patterns of the occupants, in a total evacuation scenario and to demonstrate that they are able to enter the fully protected zone in 2.5 minutes.

4   Simulation by Deterministic Fire Engineering Approach

Two scenarios are examined:

- Scenario A — evacuation with 3 staircases.
- Scenario B — evacuation with the middle stair closed.

Figure 9.2   Flow Chart Showing the Approach of the Study

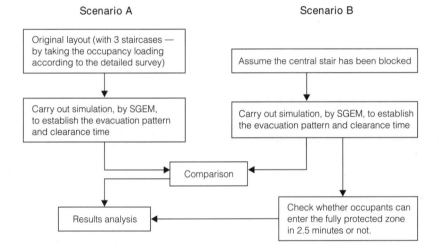

In order to examine the evacuation pattern of the two scenarios, an evacuation program — Spatial Grid Evacuation Model (SGEM) (Lo and Fang, 2000) has been adopted. The parameters adopted for the simulation are as follows:

- Simulation time step = 0.2s
- Occupancy loading taken from Table 1 of MoE Code, 96
- The reference horizontal free movement speed (v_{H0}) = 1.3 m/s
- The reference downward free movement speed (v_{V0}) = 1.1 m/s
- The actual speed of each individual is given by the equation:
  $u_i = 112\,\rho^4 - 380\,\rho^3 + 434\,\rho^2 - 217\,\rho + 57\ (m/min)$ in that $\rho = D_i \cdot f$ and $0 < f < 0.92$, D_i is the density of the zone of the evacuee.

The SGEM has been used to study the flow pattern and overall clearance time of the two scenarios in the residential block. Table 9.1 shows the simulation results.

Table 9.1   Simulation (total population =1,080 p)

| | Flow Time with 3 stairs (Scenario A) | | Flow Time with 1 stair closed (Scenario B) | |
Floor	Maximum clearance time for storey (sec)	Time for last person to enter staircase(s) (sec)	Maximum clearance time for storey (sec)	Time for last person to enter staircase* (sec)
12/F		48		63
11/F	61	49	75	64
10/F	75	46	91	63
9/F	92	58	107	78
8/F	109	52	128	66
7/F	122	46	154	67
6/F	140	48	192	68
5/F	165	49	236	68
4/F	196	48	291	64
3/F	223	46	336	63
2/F	251	46	374	64
1/F	270	47	406	63
G/F	286		425	

* Time required for evacuees to move to the other stairs not considered
Remarks: i)   The time shown on the table is an average of four simulation results.
            ii) The initial positions of the occupants were randomly generated.

It has been shown in Table 9.1 that the overall clearance time for the original layout (286s) is shorter than that of the situation where the central stair has been closed for repair purpose (425s). Nevertheless, the simulation results indicate that the clearance time per floor will still be within the 2.5 minutes (150s) notional period stipulated in the MoE Code, 1996 (maximum time = 78s at 9/F). The performance is within the acceptance criterion in the MoE Code, 96.

In order to carry out the evacuation study including the interacting effect of hot gases developed from the fire, a computer simulation on the dispersion of hot gases was carried out together with the evacuation model for comparison. A staircase is assumed to be blocked whenever hot gases from fire reach the staircase. Initially, both staircases in the left and right wings of the buildings are assumed to be available for evacuation. When the hot gases developed from the fire reach one of the staircases, this staircase is assumed to be untenable and not available for evacuation. The evacuees are then assumed to seek escape through the other staircase only.

Since this approach involves the development of a fire, a fire profile should be selected for estimating the development of the hot gases. The fire profile, as shown in Figure 9.3, in the form of a heat-release-rate against time, extracted from National Institute of Standards and Technology (NIST), was applied to the model. This is the result in test no. 5 of the NIST report — SBSIR85-2998 when the total heat release is maximal:

> The furnishing arrangement representative of those in the USA National Park Service lodging facilities was evaluated for its open burn (free burn) characteristics. The arrangement consisted of a double bed with a wooden headboard and one wooden night table. The proximity of a wall and the effect of a room on the combustion of the same arrangement were examined. The wall finish materials were gypsum board and plywood.

Figure 9.3   Fire Profile Used for Simulation by Hazard-I

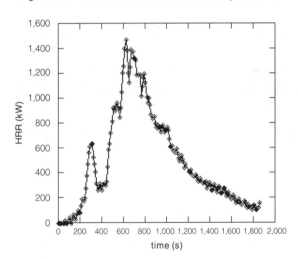

The headroom of all floor areas was assumed to be 2.5m. There were 220mm existing structural down-strands, as shown in Figure 9.1. Two new down-strands were also provided, of which the soffits were 2.0m from the finished floor level, from a reservoir for delaying smoke dispersion. The floor plan of the typical floor was divided into 3 compartments as arrowed on Figure 9.1. The fire was ignited at the centre of U1.

The dispersal of hot gases from the fire was predicted using a fire simulation software Hazard-I, developed by NIST, US Government. This is a zone model in which the distribution of energy and mass throughout the rooms included in the simulation was done in the model FAST. The basic assumption of such models was that each compartment could be divided into two or more zones, each of which was internally uniform in temperature and composition. All compartments had two zones except the fire room, which had an additional zone for the fire plume. The boundary between the two layers in a compartment is called the interface. It has generally been observed that in the spaces close to a fire, buoyantly stratified layers will be formed. While in an experiment the temperature can be seen to vary within a given layer, these variations are small compared with the temperature difference between the layers.

Figure 9.4 shows the smoke layer development simulated by Hazard-I. It can be observed that the layer of hot gases at U3 (the right wing corridor) was always higher than 2.0m. As the soffit of the new down-strand at the entrance of the right wing staircase was 2000mm above FFL, we can assume that no smoke was able to disperse into the right staircase and that it provided a safe escape route for evacuation. However, the layer of hot gases reaches the soffit of the down-strand at the entrance of the left staircase in 165 seconds. It is thus assumed that the smoke will disperse into the left staircase 165 seconds after the fire ignites, and the staircase will not be safe for evacuation. The evacuees will then move to the other stair for escape.

Figure 9.4   Development of Hot Gas Layers for Fire Ignition at U1

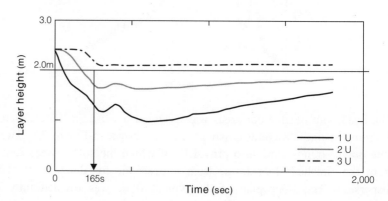

Table 9.2 *Simulation with the Left Stair Blocked by Smoke at 165s*

Floor	Flow Time with 3 stairs (Scenario A)		Flow Time with 1 stair closed (Scenario B)	
	Maximum clearance time for storey (sec)	Time for last person to enter staircase(s) (sec)	Maximum clearance time for storey (sec)	Time for last person to enter staircase* (sec)
12/F		48		63
11/F	61	49	75	64
10/F	75	46	91	63
9/F	92	58	107	78
8/F	109	52	128	66
7/F	122	46	154	67
6/F	140	48	(left stair blocked by smoke)	
6/F	(left stair blocked by smoke)		250	68*
5/F	215	49*	266	68
4/F	230	48	301	64
3/F	246	46	345	63
2/F	262	46	402	64
1/F	280	47	466	63
G/F	301		515	

* time required for evacuees to move to the other stairs not considered.

When the smoke entering the left stair is not considered, we can define the change of safety factor (F_S) with respect to the 2.5 minutes requirement in the MoE Code, 96 as follows:

a) *when 3 stairs are used for escape:* $F_S = \dfrac{150}{58} = 2 \cdot 59$

b) *when 1 stair blocked for repair:* $F_S = \dfrac{150}{78} = 1 \cdot 92$

The reduction in F_S is about 26%.

When the smoke entering the left stair is considered, the FS may be regarded as the same, whereas the safety factor with respect to the overall clearance time of the building will be reduced to $\dfrac{515 - 301}{301} = 71 \cdot 1\%$.

In order to assist the occupants to find a way (or ways) to escape, additional signage should be provided at various critical locations. Directional signs should be painted in the corridors at low levels in order to direct evacuees to a suitable stairway. Adequate fire separation between the construction areas and the main occupancy of the building should also be provided. The management company of the residential building must undertake to improve the management level for the building, including providing additional training to the staff.

Table 9.3 Assessment of the Scenario in which the Mid-stair Is Not Closed

Fire Safety Attributes		Overall Weighting	Score*	Weighted Score		Works Done
Means of Escape	Number of Exits	0.046	3	0.138	0.731	Mid-stair closed and exit number, travel distance and total width of exit routes affected
	Travel/Direct Distance	0.061	3	0.183		
	Width of Exit Route	0.040	3	0.120		
	Exit Signs	0.067	2	0.134		
	Emergency Lighting	0.078	2	0.156		
	Refuge Floor	0.023	0	0.000		
Fire Resistant Construction	Compartmentation	0.034	4	0.136	0.574	—
	Fire Rated Door	0.047	2	0.094		
	FR Walls/Floors	0.029	4	0.116		
	External Flame/Smoke Spread	0.042	3	0.126		
	Internal Flame/Smoke Spread	0.034	3	0.102		
Means of Access for Fire Fighting and Rescue	FF and R Stair	0.023	3	0.069	0.174	—
	Firemen's Lift	0.017	3	0.051		
	Smoke Vent	0.015	0	0.000		
	EVA	0.020	0	0.000		
	Ventilated Lobby	0.018	3	0.054		
Fire Services Installations	Fire Hydrant	0.016	3	0.048	0.303	—
	Fire Alarm	0.027	3	0.081		
	Hose Reel	0.033	3	0.099		
	Fire Blanket	0.016	0	0.000		
	Auto-detection	0.028	0	0.000		
	Sprinkler System	0.071	0	0.000		
	Portable FE	0.025	3	0.075		
Building Management and Maintenance	Maintenance Level	0.072	3	0.216	0.481	—
	Staff Training	0.065	2	0.130		
	Maintenance Level	0.045	3	0.135		
Total				2.263	2.263	—

*Score, based on a 5-grade measurement, in which 1 is very poor, 2 is poor, 3 is moderate, 4 is good, 5 is very good.

Table 9.4   Assessment for the Scenario in which the Mid-stair Is Closed

Fire Safety Attributes		Overall Weighting	Score*	Weighted Score		Works Done
Means of Escape	Number of Exits	0.046	2	0.092	0.874	Additional exit signs and low level lighting/ emergency lighting added
	Travel/Direct Distance	0.061	2	0.122		
	Width of Exit Route	0.040	2	0.080		
	Exit Signs	0.067	4	0.268		
	Emergency Lighting	0.078	4	0.312		
	Refuge Floor	0.023	0	0.000		
Fire Resistant Construction	Compartmentation	0.034	4	0.136	0.608	Down-strand added to create a smoke reservoir effect at U2 and U3
	Fire Rated Door	0.047	2	0.094		
	FR Walls/Floors	0.029	4	0.116		
	External Flame/Smoke Spread	0.042	3	0.126		
	Internal Flame/Smoke Spread	0.034	4	0.136		
Means of Access for Fire Fighting and Rescue	FF and R Stair	0.023	3	0.069	0.174	—
	Firemen's Lift	0.017	3	0.051		
	Smoke Vent	0.015	0	0.000		
	EVA	0.020	0	0.000		
	Ventilated Lobby	0.018	3	0.054		
Fire Services Installations	Fire Hydrant	0.016	3	0.048	0.303	—
	Fire Alarm	0.027	3	0.081		
	Hose Reel	0.033	3	0.099		
	Fire Blanket	0.016	0	0.000		
	Auto-detection	0.028	0	0.000		
	Sprinkler System	0.071	0	0.000		
	Portable FE	0.025	3	0.075		
Building Management and Maintenance	Maintenance Level	0.072	4	0.288	0.618	Improve management and staff training
	Staff Training	0.065	3	0.195		
	Maintenance Level	0.045	3	0.135		
Total				2.577	2.577	

5 Evaluation by Fire Safety Ranking Approach

In order to provide a simple and clear illustration, a direct-point auditing approach is used for the ranking evaluation. Suppose S, the total evaluation, for a building is expressed as:

$$S = \sum_{i=1}^{n} w_i x_i \tag{1}$$

where is a parameter contributing to fire safety and is the corresponding weighting. Naturally, the weights are fixed and generally they are independent of risk assessment (Magnusson, 1998).

A hierarchy of fire safety attributes (Lo *et al.*, 2000) and the evaluations are shown in Tables 10.3 and 10.4. In this study, a subjective evaluation, on the basis of a 5-grade approach, has been used for measuring the performance of each attribute. In real situations, a checklist providing detailed evaluation criteria for each attribute should be used for the assessment.

6 Summary

According to the deterministic approach, the proposed temporary closure of the mid-stair can still meet the 2.5 minutes standard stipulated in the MOE Code. However, this approach cannot easily provide a quantitative assessment on the effect of the additional works, such as the improved management and staff training, signage and emergency lighting. It can only show the escape and smoke spread pattern in different scenarios for the purpose of comparison. In contrast, the fire safety ranking method adopts a holistic evaluation approach and can also provide a semi-quantitative approach for comparison purposes. The example shows that if we improve the management and staff training as well as the signage and emergency lighting levels, then the fire safety level of the proposed layout appears to be not lower than the existing situation (2.577 > 2.263). Although trade-off in some instances may not be allowed, if strict compliance with the codes' requirements cannot be achieved, we could then use the ranking method to evaluate the situation and provide indications for decision-making.

References

1. Lo, S. M. and Fang, Z. 2000. A Spatial-Grid Evacuation Model for Buildings. *Journal of Fire Science*, 18(5), 376–394.

2. Lo, S. M., Lam, K. C. and Yuen, K. K. 2000. Views of Building Surveyors and Building Services Engineers on Priority Setting of Fire Safety Attributes for Building Maintenance. *Facilities*, 18(13/14), 513–523.

3. Magnusson, S. E. and Rantatalo, T. 1998. Risk Assessment of Timber Frame Multi-Storey Apartment Buildings — Proposal for a Comprehensive Fire Safety Evaluation Procedure. *Internal Report 7004*, Department of Fire Safety Engineering, Lund University.

4. National Fire Protection Association. 1995. NFPA101A, *Guide on Alternative Approaches to Life Safety*. Quincy, MA

5. Watts J. M. Jr. 1995. Fire Risk Ranking. SFPE Handbook of Fire Prevention Engineering. 12–5:12–26. In P. J. Di Nenno *et al.* (eds) *National Fire Protection Association*. Quincy, MA.

Contributors

Kenneth J. K. CHAN
BSc (Leics) FHKIS FRICS FIBC
FBEng FPFM CFM CFMJ
AHKIArb RPS(BS) AP(S)

Trained in building development control, Mr. Kenneth Chan specializes in building maintenance, facility enhancement, property management, project development and management. His current interests are project management, strategic facility management and green building.

K. W. CHAU
BSc (Building Studies),
B Building, PhD, FRICS, FHKIS,
FCIOB

Professor K. W. Chau is Chair Professor of Real Estate and Construction and Dean of Faculty of Architecture, The University of Hong Kong. He was also the president of the International Real Estate Society (2000–01) and first president of the Asian Real Estate Society (1996–7). In 1999 he also received The University of Hong Kong Outstanding Young Researcher Award and the International Real Estate Society Achievement Award in recognition of his achievements in real estate research. Professor Chau's main areas of research are real estate investment, real estate finance and economics, construction economics, building performance assessment and performance of the real estate and construction sectors. Most of his works are empirical studies with implications for policy makers, developers and investors. He has published and presented more than 200 technical/academic papers and reports. Professor Chau also serves on the editorial boards of 13 international peer-reviewed academic journals.

Grace W. Y. CHENG
MHKIS, MRICS

Miss Grace Cheng is a qualified Building Surveyor with experience in maintenance and renovation works for hospitals and schools. She is now an Instructor in the Department of Building and Construction, City University of Hong Kong. She teaches building design, technology and building control. Her research focuses on sustainable development and Chinese architecture.

Barnabas H. K. CHUNG
MSocSc (Pub. Admin.), FRICS,
FHKIS, FBEng, FFB, FRSH,
FPFM, ACIArb, MCMI, RPS (BS)

Professor Barnabas Chung began his career in statutory building control in 1968 and was Chief Building Surveyor of the Buildings Department, Hong Kong, until his retirement in 1998. He began teaching building law in 1983 in the Department of Engineering and the Department of Architecture at the University of Hong Kong. In 2003, he completed a comprehensive review of the Buildings Ordinance and Regulations for the Buildings Department. His current research activities include: Advisor to the consultancy study on Review of Lighting and Ventilation Requirements in Buildings, Advisor to the consultancy study on Comprehensive Environmental Performance Assessment Scheme, Advisor to the consultancy study to draft a new Manuel on Barrier Free Access. He received an International award from the World Organization of Building Officials in 1996, and the Distinguished Building Surveyor Award from the Hong Kong Institute of Surveyors in 1998.

Charles C. K. HO
BSocSci, MSc

Charles Ho worked as a research assistant at the City University of Hong Kong and is now studying a Master of Philosophy in Sociology degree at the Chinese University of Hong Kong. His research interests include the emergence and the type of urban collective action using the methods of event history analysis and ethnography, urban sociology, political sociology, social movement and ethnography.

Daniel C. W. HO
BSc, MBA, PhD, FHKIS, FRICS,
FPFM, RPS (BS), Authorized
Person (Surveyors)

Dr. Daniel C. W. Ho is a qualified Building Surveyor and a Fellow of both the Hong Kong nstitute of Surveyors (HKIS) and the Royal Institution of Chartered Surveyors (RICS). He is currently Associate Professor in the Department of Real Estate & Construction, The University of Hong Kong. Dr. Ho is a Founding Member and Fellow of the Hong Kong Institute of Facility Management (HKIFM). He is currently President of the HKIFM and the co-ordinator of the MSc (Real Estate) majoring in Facility Management program offered by the Department of Real Estate and Construction, The University of Hong Kong. His research interests and expertise are in facility performance assessment, development control and facility management.

P. K. KAM
BSocSci (Hons) (Social Work),
MSc (Advanced Social Work
Studies), PhD, RSW

Dr. Kam is Associate Professor in the Department of Applied Social Studies at the City University of Hong Kong. He teaches social work courses, and his research interests are in urban renewal and housing issues, community development, empowerment practice, social gerontology and social work education His recent published books include: Analysis of housing policy in Hong Kong (in Chinese) (1996) (co-editor). Skills in community work (in Chinese) (1997) (co-editor); Community work theory and practice (in Chinese) (1994) (co-editor). He is the winner of Teaching Excellence Award (1994) and the Faculty of Humanities and Social Sciences (FHS) Contribution to Learning Award (1999), City University of Hong Kong.

Lawrence W. C. LAI
BSocSc (Econ); MTCP,
MSocSc (Econ), LLB, PhD,
MHKIP, MRAPI, MCILT,
MIAAEM, FPFM, RPP

Professor Lawrence Lai is experienced in town and country planning, planning appeals, sustainable development. His research focuses on the study of property rights aspects of planning and development, planning appeals and enforcement, heritage planning, aquaculture. His recent key publications include Property Rights, Planning and Markets (with Chris Webster), Edward Elgar, Cheltenhem, 2003; and Understanding and Implementing Sustainable Development (eds. Lawrence W. C. Lai and Frank T. Lorne), Nove Science, New York, 2003.

S. M. LO
PGDip, MSc, PhD, FRICS,
FHKIS, Authorized Person

Dr. S. M. Lo is Associate Professor in the Department of Building and Construction, City University of Hong Kong. He is also a Visiting Professor at the Xi'an Jiaotong University and Shenyang Institute of Architecture and Civil Engineering, China. He is a chartered Building Surveyor and Authorized Person. He was a building official in the Hong Kong Government responsible for scrutinizing building development proposal and drafting legislative proposals and codes of practices. In the past ten years, he has served many statutory committees. He was the secretary of the Working Party on the Review of Means of Escape Code, Member of the Contractors' Registration Committee, Member of the Fire Safety Committee and etc. Apart from professional practice, he has actively participated in many research projects. His main research interests include building design, fire safety engineering and decision support system. Currently, he holds many research grants including 6 Competitive Earmarked Research Grants from the Research Grant Council of HKSAR for studying evacuation, fire risk analysis, decision support system, wayfinding modelling, intelligent understanding of CAD plans, human behaviour in fire, fire modelling and effect of fire legislation on the society. He has over 150 publications including about half in refereed journals.

Sik Hung NG
BSocSc (HKU), PhD (Bristol)

Dr. Ng held a Chair in Psychology at the Victoria University of Wellington before he took up his current position as Professor (Chair) in Social Psychology at the City University of Hong Kong. His research focuses on positive ageing, inter-generational relations and the bicultural self. He is a Fellow of the Royal Society of New Zealand and President-Elect of the International Association of Language and Social Psychology.

Raymond W. PONG
BA (Hons), PhD

Dr. Pong is Research Director of the Centre for Rural and Northern Health Research and a full professor at Laurentian University (Ontario, Canada). He obtained his B.A. (Hons.) from the University of Hong Kong and Ph.D. in sociology from the University of Alberta (Canada). His research interests include the health workforce, health policy, rural health, gerontology and program evaluation. His research has been funded by such granting agencies as the Canadian Institute of Health Research, the Canadian Health Services Research Foundation and the Richard Ivey Foundation. He has done consultancy work for the Commission on the Future of Health Care in Canada and the World Health Organization.

C. M. TAM
MSc, PhD, FCIOB, MHKIE,
FHKICM

Dr. Tam is currently Associate Professor and Associate Head of the Department of Building & Construction, City University of Hong Kong. He had worked years in the building industry before he joined the City University in 1986. In 1984, he finished a M.Sc. program in Loughborough University in the UK. Then he worked as a project manager and participated in some of the earliest Hong Kong invested projects in hotels and factories in China. He obtained a Ph.D. in the same university in 1993. He has been serving as leaders of several teaching programs. His main research interests include safety and quality management, use of information technology in construction and education, productivity studies, BOT and procurement systems, etc.

S. K. WONG
BSc (Surveying), PhD (HK), PFM

Dr. S. K. Wong is a Lecturer in the Department of Real Estate and Construction at the University of Hong Kong. He is a Professional Facility Manager and his research interests include property price indexes, real estate economics and finance, and facility benchmarking. His most recent publications includes "Hedonic price modelling of environmental attributes: a review of the literature and a Hong Kong case study" (2003), in Understanding and Implementing Sustainable Development (eds. L. W. C. Lai and F. Lorne), Nova Science, 87–110. (with K. W. Chau, C. Y. Yiu and L. W. C. Lai)

W. S. WONG

BA (AS), BArch, FHKIA, RIBA, ARAIA, Authorized Person

Dr. W. S. Wong is an Associate Professor of the Department of Architecture, The University of Hong Kong. He is also a registered Architect and an Authorized Person.

C. Y. YIU

BSc (Surveying), MPhil (HK), PhD (HK), MRICS, MHKIS, PFM

Dr. C. Y. Yiu is an Assistant Professor in the Department of Building and Real Estate, The Hong Kong Polytechnic University. He is a qualified Building Surveyor and a Professional Facility Manager. His research interests include building maintenance, facilities management, as well as economics and finance in property and construction. In 1998, he obtained the first prize in the Hong Kong Economic Journal Thesis Competition and the second prize of the Dr. Lee Shiu Prizes for Essays and Theses on the Development of Hong Kong since World War II.

Richard K. K. YUEN

PhD

Dr. Richard K. K. Yuen obtained his PhD in Fire Dynamics and Engineering from the University of New South Wales, Australia. He joined the Department of Building and Construction at the City University of Hong Kong in 1990. He has worked for General Electric Company (Hong Kong) Limited and Hong Kong Electric before taking up the lecturership in the Department of Building Services Engineering at the Hong Kong Polytechnic University. He is a Chartered and Registered Professional Engineer and member of various professional bodies including the Hong Kong Institute of Engineers, the Chartered Institute of Building Services Engineers and Institute of Australian Engineers. He teaches Built Environment, Fire Engineering and Building Services Engineering. Dr. Yuen is the Programme Leader of the BEng (Honours) in Building Services Engineering (Full-time) degree programme. His research interests include fire safety and engineering, pyrolysis and combustion, applications of computational fluid dynamics, neural network modelling applications in fire engineering, building energy conservation, lighting and ventilation, HVAC systems and indoor air quality.

Index